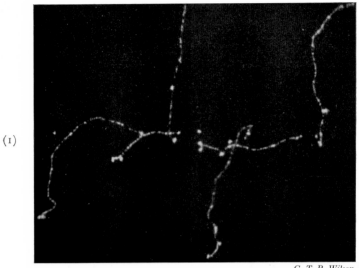

C. T. R. Wilson

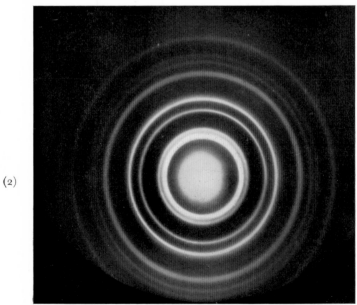

G. P. Thomson

ELECTRONS BEHAVING AS (1) *Particles*, in passing through a gas;
(2) *Waves*, in passing through a thin metal film (p. 199)

The
New Background of Science

by

SIR JAMES JEANS

M.A., D.Sc., Sc.D.
LL.D., F.R.S.

CAMBRIDGE
AT THE UNIVERSITY PRESS
1933

PRINTED IN GREAT BRITAIN

CONTENTS

Electrons behaving as (1) *Particles*, (2) *Waves*   *frontispiece*

Preface   *page* vii

Chapter I   The Approach to the External World   1

      II   The Methods of Science   45

      III The Framework of the External World—
            Space and Time   70

      IV   Mechanism   112

      V   The Texture of the External World—
            Matter and Radiation   147

      VI   Wave-Mechanics   194

      VII Indeterminacy   231

      VIII Events   262

Index   299

PREFACE

After undergoing a succession of kaleidoscopic changes, theoretical physics appears to have attained a state of comparative quiescence, in which there is fairly general agreement about essentials. In the following pages I have tried to depict the present situation in broad outline and in the simplest possible terms. I have drawn my picture against a roughly sketched background of rudimentary philosophy—the philosophy of a scientist, not of a metaphysician—because I believe, in common with most scientific workers, that without a background of this kind we can neither see our new knowledge as a consistent whole, nor appreciate its significance to the full. Statements made without reference to such a background—as, for instance, that "an electron consists of waves of probability" or that "the principle of indeterminacy shews that nature is not deterministic"—can convey at best only a minute fraction of the truth.

I have tried to exhibit the new knowledge in such a way that every reader can form his own judgment as to its philosophical implications. There is room for much legitimate difference of opinion as to what precisely these are; yet few, I think, will be found to doubt that some re-orientation of scientific thought is called for. I have not suppressed my own view that the final direction of change will probably be away from the materialism and strict determinism which characterised nineteenth-century physics, towards something which will accord better with our

everyday experience. This part of my work may be regarded as an amplification and clarification of parts of my
earlier small book, *The Mysterious Universe.*

I have hoped that the present book may serve a serious
scientific purpose, and prove of interest and value both to
students of physics and to other more general readers.
Unhappily I found it impossible to attain the necessary
precision of thought and statement without occasionally
using a few mathematical symbols and formulae; at the
same time I have tried to arrange that the general purport
of these shall be made clear to the non-mathematical
reader, who will, I hope, find most of the book intelligible.

J. H. JEANS

DORKING
January 19th, 1933

THE APPROACH TO THE EXTERNAL WORLD

Twentieth-Century Physics

A century which has run less than a third of its course has already witnessed two great upheavals in physical science. These are associated with the words Relativity and Quanta, and have forced the physicist of to-day to view nature against a background of ideas which is very different from that of his nineteenth-century predecessor.

The latter thought of nature as an assemblage of objects located in space and continually changing with the passage of time. It was something entirely detached from, and external to, himself; something which he could study and explore from a distance as the astronomer studies the surface of the sun through his telescope, or the explorer the desert from his aeroplane. He thought of the apparatus of his laboratory as the astronomer thinks of his telescope, or the explorer of his field-glass; it shewed him things which were there whether he looked at them or not, which had been there before the first man appeared on earth, and would still be there after the last man had been frozen to extinction. Finally he accepted a "common-sense" view of nature, believing that there was no great difference between appearance and reality; the possibility that things were not as they seemed might provide an admirable subject for a debating society of philosophers, but was of as little practical concern to the scientist as to the farm-labourer.

Although he may not have realised it, this complex of beliefs constituted a philosophical creed in itself. No attempt was made to justify it by abstract argument; so long as it worked satisfactorily none seemed to be needed, the success of the science based upon it providing a sufficient justification. If ever it ceased to work, there would be time enough to probe its foundations and perhaps look for a new philosophy.

That time has now come. The old philosophy ceased to work at the end of the nineteenth century, and the twentieth-century physicist is hammering out a new philosophy for himself. Its essence is that he no longer sees nature as something entirely distinct from himself. Sometimes it is what he himself creates, or selects or abstracts; sometimes it is what he destroys.

In certain of its aspects, which are revealed by the new theory of quanta, nature is something which is destroyed by observation. It is no longer a desert which we explore from the detached position of an aeroplane; we can only explore it by tramping over it, and we raise clouds of dust at every step. Trying to observe the inner workings of an atom is like plucking the wings off a butterfly to see how it flies, or like taking poison to discover the consequences. Each observation destroys the bit of the universe observed, and so supplies knowledge only of a universe which has already become past history.

In certain other aspects, especially its spatio-temporal aspects as revealed by the theory of relativity, nature is like a rainbow. The ancient Hebrew—the analogue of the nineteenth-century physicist—saw the rainbow as an objective structure set in the heavens for all men to behold, the token of a covenant between God and man, and as objective as the signature to a cheque. We now know that the objective rainbow is an illusion. Raindrops break sunlight up into rays of many colours, and the coloured rays which enter any man's eyes form the rainbow he sees; but as the rays which enter one man's eyes can never enter those of a second man, no two men can ever see the same rainbow. Each man's rainbow is a selection of his own eyes, a subjective selection from an objective reality which is not a rainbow at all. And it is the same with the nature which each man sees.

Again, just as a man's rainbow follows him about as he moves round the country-side, so nature follows us about. At whatever speed we move, we find nature adjusting itself to our motion, so that this motion makes no difference to its laws.

Yet the analogy fails in one respect. A rainbow will disclose our own motion to us by the speed with which it moves against a background of distant forests and hills, but physical science can find no such background for nature. The whole of nature appears to follow us about.

Imperfect though these analogies are, they will shew that the physicist of to-day must needs have some acquaintance with ideas which used to be considered the exclusive preserve of metaphysics.

One of the foremost workers in modern theoretical physics, Professor Heisenberg of Leipzig, has described the present situation in the following words:*

"With the advent of Einstein's relativity theory it was necessary for the first time to recognize that the physical world differed from the ideal world conceived in terms

* *The Physical Principles of the Quantum Theory* (Univ. of Chicago Press, 1930), p. 62.

of everyday experience.... The experimental material resulting from modern refinements in experimental technique necessitated the revision of old ideas and the acquirement of new ones, but as the mind is always slow to adjust itself to an extended range of experience and concepts, the relativity theory seemed at first repellantly abstract. None the less, the simplicity of its solution for a vexatious problem has gained it universal acceptance. As is clear from what has been said, the resolution of the paradoxes of atomic physics can be accomplished only by further renunciation of old and cherished ideas....

"To mold our thoughts and language to agree with the observed facts of atomic physics is a very difficult task, as it was in the case of the relativity theory. In the case of the latter, it proved advantageous to return to the older philosophical discussions of the problems of space and time. In the same way it is now profitable to review the fundamental discussions, so important for epistemology, of the difficulty of separating the subjective and objective aspects of the world. Many of the abstractions that are characteristic of modern theoretical physics are to be found discussed in the philosophy of past centuries. At that time these abstractions could be disregarded as mere mental exercises by those scientists whose only concern was with reality, but to-day we are compelled by the refinements of experimental art to consider them seriously".

This is not meant in any way to suggest that an objective nature does not exist, but merely that it is at present beyond our purview. We can only see nature blurred by the clouds of dust we ourselves make; we can still only see the rainbow, but a sun of some sort must exist to produce the light by which we see it.

Writing in 1899,* F. H. Bradley proposed to define the nature of metaphysics as

* *Appearance and Reality*, p. 261.

"the bare physical world, that region which forms the object of purely physical science, and appears to fall outside of all mind. Abstract everything psychical, and then the remainder of existence will be Nature".

A few lines farther on, he brings us to the crux of the present situation in physical science when he writes:

"We sometimes forget that this world [of nature], in the mental history of each of us, once had no existence. There was a time when the separation of the outer world, as a thing real apart from our feeling, had not even been begun. The physical world, whether it exists independently or not, is, for each of us, an abstraction from the entire reality".

A nineteenth-century physicist, reading this, would have identified the "time when the separation of the outer world had not even been begun" with a few days in his extreme infancy, and would little suspect that he, a scientist of mature years, had not yet effected the separation completely. It was left for twentieth-century physics under the lead of Einstein, Bohr and Heisenberg to discover how large a subjective tinge entered into the nineteenth-century description of nature; recognising this, it tries to discard our human spectacles and study the objective reality that lies beyond. Only in this way has it proved possible to give a consistent description of nature. Thus the history of physical science in the twentieth century is one of a progressive emancipation from the purely human angle of vision.

The physicist who can discard his human spectacles, and can see clearly in the strange new light which then assails his eyes, finds himself living in an unfamiliar world, which even his immediate predecessors would probably fail to recognise.

We must now try to explain how this change of thought has come about, examine its implications, and describe, in so far as this is possible, the new world of twentieth-century physics.

The World of Sense-impressions

We may properly approach this world by imagining the entry into life of a child endowed with consciousness, with a mind capable of experiencing sensations and desires, and with a capacity for thought.

At first it has no consciousness except of its own existence; no knowledge of an outer world of nature, as something distinct from and clearly separated from itself, its thoughts and its sensations; no past experiences to form a background to its thoughts or with which to compare its present sensations. Gradually the passage of time provides past experiences, which memory fixes in its mind to form the needed background. It begins to view its sensations against this background, and discovers that they continually change. They fall into the two categories of pleasurable sensations, which it desires to increase, repeat or perpetuate, and painful sensations, which it desires to diminish or avoid. Soon it makes the melancholy discovery that it cannot by its own volition make all its sensations pleasurable; it finds that it has needs, such as for food and warmth; when these are not adequately satisfied, its sensations are less pleasurable than when they were. These needs introduce it to the hard facts of life, for it finds they can only be satisfied from outside itself. Definite acts, such as sucking sugar or running a pin into its hand, produce still more acute sensations of pleasure or pain; the materials for these sensations, the sugar or the pin-point, also come to it from outside.

From such experiences the child infers the existence of an environment which is not part of itself—in brief, of an external world. It has every inducement to try to understand the workings of this external world, in which it believes all physical pains and pleasures to originate. It soon learns, when burnt, to dread the fire; once bitten, it is twice shy. Through such experiences, it finds law and order in the external world, and discovers the principle which it will describe in later years as the "uniformity of nature"—like causes produce like effects. Finally, in its efforts to understand the external world, it begins tentatively to endow this world with certain qualities, properties and occupants. The inference that an external world exists obviously stands on a higher level of probability than the conjecture that any special qualities, properties or occupants are associated with it.

For the child has definite knowledge only of the sensations in its own mind. If these originated solely in its own mind, it could choose to make them all pleasurable; since it cannot do this, it is on fairly safe ground in supposing that something external must exist to produce and control these sensations. On the other hand, the nature of this something can never be more than guessed. The child will never be able to test the absolute truth of its conjectures; the most stringent test available is that of their consistency with one another and with the phenomena which they attribute to the external world. Such a test may disprove, but can never prove.

Throughout its whole life, the child will assume that an external world exists, and will make conjectures with a view to understanding its workings. When it does this in a logical and systematic manner, we call it a scientist.

The child's sensations reach its mind through five channels, which we call the five senses—sight, hearing, smell, taste and touch. These all function in similar ways. Something external produces an impression on some part of the body—the retina, the ear-drum, the nostrils, the palate or the skin—and this impression is transmitted along a complicated nervous system to the brain. Up to this stage the impression has been conveyed by atomic changes, but it now crosses what we may describe as the "mind-body" bridge, and when it appears on the other side, it is as a mental sensation, accompanied by such mental attributes as pleasure or pain, satisfaction or irritation, ecstasy or despair.

The nerves may be compared to a number of telephone wires transmitting electric currents into a prison-cell, which suitable instruments subsequently metamorphose into messages of sound, television, etc. The child is a prisoner inside the cell, and is doomed to remain a prisoner all its life. It can have no knowledge of the outer world except through the messages received over the wires. These may give truthful reports of the events occurring outside the prison cell, but its occupant will only be able to interpret them in terms of the contents of its cell, which consist of thoughts and sensations. A mind which is directly acquainted only with thoughts and sensations may be as little able to form a true picture of an outer world as a blind man is able to understand the beauty of a sunset or a deaf man to grasp the meaning of a symphony. Even a superior being coming direct from the outer world might still be unable to explain its nature to the prisoner, for the simple reason that they would have no common language in which to converse. Nevertheless, from the fragmentary messages which his senses send to him over his nerves, the prisoner may attempt to form a consistent picture of the external world for himself, in terms of the concepts with which his mind is familiar. Science merely attempts to build up such a picture in a systematic, organised way.

The External World

The first messages which the child receives from its senses teach it to regard the external world as a collection of objects, each possessing a certain degree of persistence or continuity in time. It soon finds that these fall into distinct categories. First come other human beings, similar to itself except for differences in age, size and other characteristics. There are also animals, birds, fishes and insects, then plants and trees, and finally objects which consist of in-animate matter.

The child's mind is not only occupied by its sensations but also by its volitions, which are desires to increase or diminish particular sensations according as it finds them pleasurable or the reverse. Having discovered that its sensations come to it from the disposition of the objects of the external world, it would like to alter this disposition, so as to avoid pain and increase pleasure. It finds, or thinks it finds, that it is possessed of a will-power, through which it may at least try to effect the changes it desires.

It soon discovers an essential difference between animate

and inanimate objects. After a little experience, it can catch a rolling marble without difficulty, because this has no will-power to set in opposition to its own, but as soon as it tries to catch a crawling fly or a crawling wasp, it becomes conscious of an opposing will-power; the fly tries to avoid capture, the wasp resents capture. Finally it finds that other children have a will-power of the same kind as its own. As it believes its will-power to emanate from its mind, it infers that the external world is controlled in part by minds other than its own, but similar to its own; it concludes that it is not the only mind in the universe.

When it establishes contact with these other minds, it learns that they experience sensations and desires similar to its own; not only are they endowed with similar senses but also, most important of all, they perceive objects similar to those which it perceives.

Not only are these objects similar in kind; often they are obviously identical. If I count that there are six chairs in my room, the normal event will be for my companion also to count six chairs. Repeated experiences such as these suggest that the chairs he sees are identical with those which I see. The knowledge that a chair can be perceived by a mind is extended to the knowledge that the same chair can be perceived by two minds, and we conclude that the chairs have what we may call an "objective" existence—an existence outside our individual minds. Something outside both of us, which we loosely describe as a chair, can produce in both of us the sense-impression we describe as seeing a chair. At this stage we naturally begin to inquire—"What is this object which we call a chair?" We turn to the physicist for an answer, because he has devoted his life to investigating such problems.

Matter

He tells us in the first place that all sense-impressions which come to us from the external world originate in what he calls "matter". This cannot of itself make a direct impression on our senses; such impressions are only made by physical "events" occurring in matter. Strictly speaking, we do not see the sun; we see events taking place in the sun. The sun only affects our senses because a continuous re-arrangement of electrons in the solar atoms results in the emission of light. In the same way, we do not see a chair, but the event of daylight or electric light falling on a chair. If we stumble against the chair in the dark, we do not feel the chair, but the event of a transfer of energy and momentum between the chair and our bodies.

Both chronologically and causally the act of perception starts at the end of the chain remote from the percipient— in the sun, the electric light, or the chair. We must not, for instance, compare the act of vision, as Descartes did, to a poking about in space, as a blind man pokes about with a stick; the object is the starting-point, not the terminus, of an act of perception.

A mental impression may be produced either by the activities of the mind itself, as when I dream, or by external events which originate in matter and subsequently operate on my mind through my senses. When many of us experience the same, or very similar, mental impressions, we usually attribute them to external events. When only one person receives an impression, although others were equally in a position to do so if it had originated in external events, we may safely attribute it to the activities of the percipient's own mind, stimulated possibly by events in his body, as with the nightmares of the man who has dined too well, or the waking illusions of the man who has drunk too well.

Thus matter may be defined as that which is capable of originating objective sensations—sensations which can be perceived by anyone who is suitably conditioned to receive them—as, for instance, by sending rays of light into our eyes. The chairs in my room are material because my companion and I can both see them if we look in the proper direction with our eyes open. But if he claims to see red snakes or pink rats which I cannot see when I look where he directs me to look, I shall conclude that his sensations are peculiar to himself; the supposed snakes and rats are creations solely of his imagination, and do not consist of matter. For practical purposes, the test of the photographic plate is usually taken to be final. A hundred people may say they see an Indian climbing up a rope into heaven, but if a suitably exposed photographic plate shews no image of the Indian and his rope, we refuse to classify these as material.

In our less reflecting moments we are apt to claim a very intimate acquaintance with matter. Reflection shews through how many intervening stages our knowledge of it must come—matter, events, effect on our senses, travel along our nerves, passage over the mind-body bridge— before it reaches our minds. For this reason the matter in which events originate may often be very different from the matter we think we see or hear or feel—all magic, conjuring and unconscious self-deception rest on the possibility of this distinction. We may see or photograph a rainbow, but the light by which we do this does not originate in the rainbow we think we see; it originates in the sun, whose rays are reflected into our eyes or our camera by the drops of rain which make the rainbow. We could photograph a ghost if this consisted of moonlight reflected from a white curtain, but the light which affected our photographic film would not come from a disembodied spirit, but from the sun.

Primary and Secondary Qualities of Matter

Even when my companion and myself both see an unmistakably objective chair, the sensations which this produces in him will never be quite identical with those it produces in me. This may be due in part to our looking at the chair from different positions, but even if we look at it in succession from the same position, there will still be differences. My perception of the chair owes something to the chair, but something also to myself.

The philosophers, who took this question in hand before there was much exact scientific knowledge to guide them, proceeded by discussing all objects and material substances in terms of certain characteristic qualities or properties with which they were supposed to be endowed. A chair, for instance, was supposed to be possessed of hardness, brownness, squareness, and so on; sugar of hardness, sweetness and whiteness. They divided these qualities into two categories which they labelled as primary and secondary, or sometimes—with a different shade of meaning—as substantive and adjective. In brief, the secondary or adjective qualities were "sense-qualities", which made, or could make, a direct appeal to our senses. Such qualities might vary with the conditions of perception, or with the state of the senses of the percipient; sugar might look white on one occasion, but yellowish or greyish on another, when it was viewed in a different light, or by a sick man. These secondary or adjective qualities were supposed to result from certain primary or substantive qualities, which were not directly perceived in themselves, but persisted independently of the perceiver, and so also of his idiosyncrasies. These existed in their own right, even when the object was not perceived at all; they were the residue after all the secondary qualities had been stripped away, the bedrock underlying the ever-shifting sands of appearance. These primary qualities could only exist attached to some substratum or foundation of real substance.

There is no obvious *a priori* justification for dividing qualities into two sharply defined categories in this way, and neither does science know of any. And as no clear-cut division can be found in practise, there has been no general agreement as to which qualities were primary and which were secondary.

Descartes, for instance, maintained that the only primary qualities were extension in space and motion—"Give me extension and motion and I will construct the Universe". Locke, relying on Newton's teaching that an unchangeable mass was associated with every object, added mass to the list. Others have maintained that extension in space is the only primary quality, and that all the observed qualities of objects emanate from this. In a later chapter we shall see how the theory of relativity has shewn that neither mass nor motion nor extension in space can qualify as true primary qualities. They depend one and all on the special circumstances of the percipient, so that the mass, motion and size of a body are as much secondary qualities as the brownness of a chair or the whiteness of sugar. Thus the theory of relativity makes it clear that if primary qualities exist, we must commence the search for them afresh.

Long before the days of relativity, Bishop Berkeley (1685–1753) and his school of thought held that there were no primary qualities at all, or, more precisely, that there was no real distinction between primary and secondary qualities. They maintained that an object was nothing more than the sum of the impressions it made in our minds, so that it had no existence at all except in so far as it was perceived by a mind or existed in a mind; nothing had more substance than the things we see in a dream. This led to a philosophy of idealism—or mentalism, to use a more modern term—according to which all matter, as ordinarily understood, is an illusion; nothing exists in reality except mind.

Atomism

Let us now start the search for primary qualities anew, making use of our scientific knowledge of the physical structure and properties of matter.

Many Greek philosophers, from Democritus onward, had imagined matter to consist, in the last resort, of hard indivisible pellets, each of which possessed in itself all the characteristic properties of the substance. These pellets were at first called atoms (ἀ-τέμνειν, incapable of being divided), but are now known as molecules. Gold, for instance, was supposed to be hard and yellow because it consisted of hard yellow atoms; it appeared yellow, not because our eyes saw it yellow, but because it was yellow in itself. Yellowness was a primary quality of gold.

The atomic hypothesis remained little more than a philosophic speculation until the eighteenth century, when John Dalton shewed how it illuminated and explained Lavoisier's work on the foundations of chemistry. It gained still further in vigour in the second half of the nineteenth century, when Maxwell and others shewed how it gave a simple and natural explanation of many of the known properties of gases. It has now become an essential ingredient of physical science.

It is known that an object may be either a homogeneous mass of a single substance, such as water, or a combination or mixture of different substances, as for instance a cup of tea. Here the cup may consist of a single substance known to the public as china, and to science as kaolinite, while the tea inside it is a mixture of water and tea, with perhaps sugar and milk. It is found that every simple substance, such as water or china, is formed of exactly similar molecules, each of which possesses the same chemical properties as the substance as a whole. Even a small amount of the substance consists of a vast number of molecules; a china tea-cup will consist of about a hundred thousand million million million molecules of kaolinite, and can contain an even greater number of molecules of water.

Each molecule is built up of still simpler units, to which the name "atom" has now been transferred. Chemistry, which has methods for resolving all known substances into their constituent atoms, finds that all molecules are combinations of only 90 kinds of atoms, although reasons of an abstract kind suggest that two others will probably be found in time, and possibly even a few more.

The atoms themselves are in turn built up of still simpler units. There are believed to be only two kinds of these, known as protons and electrons. Both are charged with electricity, the charge on each electron being the same in amount as that on each proton but of the opposite sign; it is conventionally agreed to describe the charge on the proton as positive, and that on the electron as negative. The protons stay permanently at the centre of the atom, where, in combination with a certain number of electrons, they form the compact structure we describe as the "nucleus" of the atom. Outside this are more electrons, most of which are kept near to the nucleus by the attraction of opposite kinds of electricity for one another, although the outermost are gripped so loosely that they may easily become detached from the atom to which they belong.

Wherever an atom contains more protons than electrons, its total charge is positive, and it attracts further negative electricity to itself from outside, in the form of electrons, until the excess charge is neutralised. Only when this has occurred is the atom in its normal permanent state. Thus the normal atom must always contain just as many electrons as protons. The simplest atom of all, that of hydrogen, contains only one proton and one electron; the next simplest, that of helium, contains four electrons and four protons; the atom of oxygen contains sixteen of each, and so on.

These electrified protons and electrons form the basic

units of which all material objects are built. The physical properties of a particular substance are determined by the way in which these units, or their combinations—the atoms or molecules—are arranged.

If, for instance, these are spaced widely apart, it is easy to crush them closer together, and we say that the substance is soft or yielding. If they are already so close together that a great deal of pressure is needed to get them still closer, we say the substance is hard. Thus diamond is hard, but carbon and lamp-black, which consist of similar atoms in more open spacing, are relatively soft.

Again, the 18 protons and 18 electrons which form a molecule of water are so arranged that they do not obstruct the passage of light; hence water is colourless and transparent. On the other hand, the 258 electrons and 258 protons which form a molecule of kaolinite are arranged in such a way that very little light can pass through. As a consequence light which falls on a kaolinite surface is merely turned back—not regularly, like light reflected from a mirror, but irregularly and in all directions, like the water splashed from a wall on which a fire-hose is playing. If we look in the right direction, it is as certain that some of this light will enter our eyes, as it is that we shall get wet if we stand near the wall. White light, such as sunlight, is a mixture of lights of all colours, so that when sunlight falls on kaolinite, a mixture of lights of all colours is reflected back into our eyes, and we say that the kaolinite looks white. On the other hand, when kaolinite is illuminated by blue light, it can only reflect blue light—because there is no light of other colours for it to reflect—and so looks blue. We see that the whiteness of china in sunlight is a property of the illumination rather than of the substance itself. The same is true of other substances, such as paper and linen, which look white in sunlight; all these merely assume the colour of the light by which they are illuminated.

Other substances have distinctive colours of their own. For instance, the redness of a rose is not a mere quality of the illumination by which we see it. Its petals absorb light of all colours except red, but any red light which falls on them is splashed back and may enter our eyes. When we see the rose in ordinary sunlight, nothing enters our eyes but red light, and we say that the rose looks red. On the other hand, if it is illuminated by blue light, there is no red light to be turned back into our eyes, so that it looks colourless or black. In the same way a man who is colour-blind to red will see and describe a red rose as colourless or black, in all lights. For the rose can send no light into his eyes except red light, and this can make no impression on his mind. Thus the redness of a rose depends on three factors—a redness in the rose itself, a redness in the light by which it is illuminated, and a capacity for seeing redness on the part of the percipient.

This may seem to suggest that colour is a secondary quality of objects, because it depends on the senses of the percipient. Science is, however, possessed of a colour-scale which is entirely independent of the imperfections of human perceptions. We shall see later how light consists of waves of different lengths. In normal eyes the longest waves produce the colour-impressions we describe as various shades of red and orange, the shortest produce shades of indigo, violet and blue, while those of intermediate lengths produce shades of yellow and green; in abnormal eyes they may of course produce other impressions. Thus, although our sense-estimation of colours may be partly subjective, we can measure the exact lengths of the waves which constitute light, and so obtain a perfectly definite, perfectly precise and perfectly objective scale of colour.

A succession of waves in which crests and troughs occur at perfectly regular intervals is described as a uniform train of waves, and the distance between any two successive crests, or any two successive troughs, is known as its "wavelength". The scientist describes light as being of a perfectly pure colour, or "monochromatic", when it consists entirely of waves of one uniform wave-length. In general he will not say that light is red, except as a brief and convenient way of making a rough statement; in his more scientific moments he will speak of light of wavelength say ·00006562 cm., and in so doing will specify a precise shade of colour in a way which is perfectly objective, and is limited only in precision by the number of decimals he uses. The lights by which we ordinarily see things—sunlight, electric light, candle-light—are all mixtures of waves of different lengths. They may be specified as made up of various pure colours of light, each specified by its wave-length, combined in stated proportions. The instrument known as the spectroscope actually effects the analysis for us, dividing up any kind of light into its constituent pure colours. The simplest spectroscope of all is a drop of water; a more powerful spectroscope is formed by a multitude of drops of water, such as the dew on the grass or the shower, which break up sunlight into the many-coloured light of the rainbow. In this we see all the pure colours arranged in the order of their wavelengths—red, orange, yellow, green, blue, violet, indigo.

These two examples shew that if we still wish to divide the qualities of an object into primary and secondary, the existence and mode of arrangement of its protons and electrons must be held responsible for, and indeed constitute, the primary qualities of an object; such qualities as colour result from these in conjunction with the special circumstances under which the object is perceived. Yet, underlying every red we perceive, there is a true objective red associated with either the object we perceive or its illuminant.

Mechanism of Sense-perception

Before we can study the properties which objects possess in their own right, we must learn how to allow for the special circumstances of the perceiver and the act of perception. This makes it important to understand how external objects act on our five senses.

The senses of smell and taste are affected by direct contact. When I say I smell ammonia, I mean that molecules of ammonia are entering my nose and being absorbed by its membranes, thereby affecting certain nerves which transmit a message to my brain. This message produces the sensation I describe as smelling ammonia. The process of tasting is similar; to taste sugar I must place particles of sugar in contact with my palate; the absorption of these particles sends a message to my brain which produces the sensation I describe as a sweet taste. Touch also operates through direct contact; I do not feel an object until there is actual contact between part of it and my skin.

On the other hand, we hear distant objects without their coming into contact with our sense-organs. When I hear a bell, it is not through bits of the bell striking my ear-drums; it is through waves of sound, which the bell initiates, striking against my ear-drums. The vibrations of the bell set up vibrations in the surrounding air, and these set my ear-drums also into vibration. This produces the sensation which I describe as hearing the sound of a bell, although actually it is feeling the effect of waves of condensation and rarefaction of the air inside my ears. All sounds are heard by a similar process.

Thus three of our senses—smell, touch and taste—perceive an object by direct contact, while a fourth, hearing, perceives an object by means of the waves it excites in a medium of communication, which is usually the air. How does the fifth sense of sight operate? The obvious but superficial answer is that it operates through light falling upon a part of our bodies, the retina, which is sensitive to light; but this merely raises the further question: What is light? The story of efforts to answer this question forms a very long chapter in the history of science.

The Nature of Light

The outstanding and most superficially obvious property of light is its tendency to travel in straight lines—we all are familiar with the straight outline of the beam of a searchlight, and the straight shafts of light which the sun shoots through a hole in the clouds, and we all shield our eyes from a strong light by interposing an opaque object. This led the early scientists to suppose that light consisted of a shower of small particles, emitted from a luminous object like shot from a gun. Newton adopted this view and elaborated it in his *Corpuscular Theory of Light*; he supposed that we see the sun because it is continually throwing off little bits of itself, some of which enter our eyes—just as we smell ammonia through its continually throwing off little bits of ammonia, some of which enter our noses.

Yet it proved exceedingly hard to fit all the facts of observation into such a theory. It is found that a big object casts a shadow, inside which everything is protected from the light, just as though rays of light were like gun-shot. On the other hand, a very small object affords no such protection; the rays of light bend round it and re-unite behind, so that there is no region of perfect shadow to which the light does not penetrate at all. Now this property of bending round an obstacle is one which we associated with waves, rather than with projectiles. When a gun is fired, an intervening obstacle may save us from being hit by the shot, but it will not save us from hearing the noise of the gun. This is because sound travels in the form of waves, and waves can bend round an obstacle.

This similarity between light and sound led scientists to suppose that light, like sound, must consist of waves. Just as we hear a bell because it sends out waves of sound, so, it was thought, we see the sun or a candle-flame because they send out waves of light. This concept formed the basis of the *Undulatory Theory of Light*, which regarded light as consisting of waves. Newton, who had consistently maintained that light was of the nature of particles, opposed this theory, but after Fresnel had disposed of his objections, the theory was developed in great detail, and was found to explain all the facts which the corpuscular theory had failed to explain, as well as many other known properties of light. Throughout the greater part of the eighteenth and nineteenth centuries, no single fact was known to be in opposition to the Undulatory Theory, and it was regarded as providing a final and complete explanation of the nature of light.

It has since become clear that the explanation was neither final nor complete. We now know that there was a great amount of truth in the old corpuscular theory of light, and the corpuscular and undulatory concepts of light must be regarded as complementary rather than antithetical. Viewed in one aspect, light has all the appearance of waves; but viewed in another aspect, it has the appearance of particles—somewhat as a comb may look like either a row of points or a solid bar, when viewed from different directions.

We shall see below (pp. 163, 191) that there is a single self-consistent mathematical description of light which accounts for all its known qualities, both wave-like and particle-like. But for the moment we can only describe the nature of light by analogies.

A partial—although only partial, and in many ways misleading—analogy is provided by an ordinary swell at sea. In a sense this consists of waves, but in another sense of particles—the molecules of the sea. The analogy is misleading because sea-waves admit of an objective description which shews that, quite apart from our observation of them, they consist of waves and of particles at the same time. This is not so with light. It can be viewed so as to look like either particles or waves, but never like both. In so far as we make it assume the properties of particles, we make it shed those of waves, and vice-versa. And when we discard our human spectacles entirely, we find that light is neither waves nor particles.

In another respect, however, the analogy is a good one. We may regard the water of the sea either from a statistical, or from an individual, aspect. Statistically it consists of waves, but individually of molecules. In the same way, when light is viewed statistically, it exhibits many of the properties of waves; when viewed individually, of particles. A very intense light may be treated as consisting of waves, but we find it necessary to think of a minute amount of light as consisting of separate particles. Because nineteenth-century science did not concern itself with such minute amounts, it found the undulatory theory satisfactory; it could treat light as a continuous stream. But the minute amounts which are so important to twentieth-century science may more properly be compared to shot fired from a gun, almost exactly as the old corpuscular theory supposed. We shall discuss all this more fully in a later chapter (p. 216).

If, then, we regard light as consisting of particles, we may say we see the sun because it is firing shot at us. We have seen how the material structure of the sun consists of atoms, which are in turn built up of protons and electrons. It is, however, neither atoms nor protons nor electrons which the sun shoots off; there is a further constituent to all matter, which we call energy, without knowing in the least what it is. It may exist either associated with matter, or as "free" energy not attached to matter. Energy may pass from one piece of matter to another, but it may also break loose from matter entirely, and travel through space as free energy, when we describe it as radiation.

Photons

If we regard light as consisting of particles, we must regard the particles as consisting of energy. These particles of freely travelling energy, or bullets of radiation, are known as "photons". Each photon has associated with it a definite mathematical quantity of the nature of a length, and when this quantity has the same value for every member of a swarm of photons, the swarm as a whole is found to shew many of the properties which would be shewn by waves having this as the distance from crest to crest of successive waves. For this reason this quantity is usually described as the "wave-length" of the photons. We shall see what it means if we think of ordinary radio waves, which are of course merely a special kind of radiation, characterised by having a specially long wave-length. An average transmitting aerial sends out about 10^{32} photons* every second, each having a "wave-length" of, say, 500 metres. Only a minute fraction of this torrent of photons falls on a distant receiving aerial, and yet even this fraction consists of so many individual photons that it may be treated as a continuous stream; this stream behaves like a succession of waves of wave-length 500 metres.

Like all other forms of energy, photons possess the property of inertia or mass. For this reason they exert pressure on anything they strike, here again behaving like

* 10^{32} means the number 10000..., in which 32 zeros follow the initial digit 1. Also 10^{-32} means unity divided by 10^{32}. Thus

$$10^{-6} = 0\cdot000001.$$

shot from a gun. A regiment of men could be mown down by a sufficiently strong light just as surely as by the stream of shot from a machine gun. The sun discharges about 250 million tons of energy every minute. On the corpuscular view this consists of tiny massive bullets travelling at 186,000 miles a second. Some of these enter our eyes, and, impinging on our retinas, transfer their energy to our optic nerve, and give us the sensation we describe as seeing the sun. The filament in an electric light bulb discharges somewhat similar photons, although in this case only at the rate of a fraction of an ounce per million years. Some of these, entering our eyes directly, strike our retinas, and we see the filament; others falling on our tables and chairs are turned back from these to pass on to our retinas, and we say we see our tables and chairs by electric light. Thus seeing is similar to smelling, except that the distance is traversed by photons, which are bullets of energy, instead of by molecules, which are bullets of matter.

Yet the mechanism of sight is far more intricate, and gives us far more detailed knowledge, than that of smell. The molecules which affect our sense of smell travel over zig-zag paths as they are buffeted about by other molecules, and so reach our noses from all directions; we cannot usually say that a smell comes from a certain direction, but merely that the air is pervaded by a smell, or, at best, that the air which reaches us from a certain vague direction is so pervaded. Photons differ from molecules in that they do not interact with one another; nothing but matter can stop a photon, or deflect it from its course. Thus photons travel through empty space in straight lines, and we know the direction from which light reaches us with the utmost precision. Just as the lens of a camera arranges that all photons which come from the same direction shall be thrown on to the same point of a photographic plate, and so produce a picture of the world outside the camera, so the crystalline lens of the eye arranges that all photons which arrive in the same direction, and so come from the same object, shall fall on the same spot of the retina. In this way the light falling on our retina constitutes a sort of picture of all the objects which are affecting our vision at any instant, and we see these objects arranged in the right order relative to one another.

When we smell several objects at the same time, we are conscious of little more than an unassorted medley of smells; we speak of the smell of the East, or the smell of a ship, without being able to enumerate the separate constituent smells. It is much the same with our palates; we taste the dish rather than its separate ingredients, which are known only to the cook. Our ears do somewhat better for us. When we hear a number of sounds simultaneously, our ears analyse the resultant sound into its constituent tones of different pitch; it is in this way that we recognise individual voices and separate musical instruments, that the ordinary ear can concentrate on the voice of a companion to the exclusion of much louder sounds, and that the trained musical ear can analyse a chord into its constituent notes. But our eyes form enormously less effective analysing instruments than any of these. They can only inform us as to the direction from which light arrives, and have no capacity at all for analysing a beam of mixed light into its constituent colours.

Just as there are sounds too deep or too acute for us to hear, so there are photons which we cannot see. Some are of too short a wave-length to be seen; none of our sense-organs apprehends these directly, although they may make painful burns on our skin. Others are of too long a wave-length to be seen; many of these represent heat rather than light, and their impact on our skin tells us of the warmth of the sun or the heat of a fire. We see, then, that our sense of touch can perceive photons as well as material objects.

The Outer World

Thus all our five senses act in essentially the same way; something ponderable from the outer world—something of which we can say that its weight is so-and-so—comes into contact with our sense-organs. We feel, taste and smell sugar by the direct contact of our skin and membranes with small particles of sugar. We hear a bell when particles of air, set into rhythmic motion by the bell, strike upon our ear-drums. We see the sun by certain of the photons which it emits striking our retina. We feel its heat by certain other of its photons impinging on our skin.

In general, then, we may say that we experience the outer world through small samples of it coming into contact with our sense-organs. The outer world consists of matter and energy; samples of this outer world consist of molecules and photons.

Yet not all samples of the outer world affect our sense-organs. Our ear-drums are affected by ten octaves, at most, out of the endless range of sounds which occur in nature; by far the greater number of air-vibrations make no effect on them. Our eyes are even more selective; speaking in terms of the Undulatory Theory of Light, these are sensitive to only about one octave out of the almost infinite number which occur in nature.

It is often maintained that, as we cannot experience the whole of nature, we can never hope to understand it. Animals exist whose senses are very different from our own; bats and cats are said to hear and see different things from ourselves, while dogs obviously smell different things. The world must seem very different to them. In the same way, if the sensitiveness of our organs were shifted to different ranges, or if we were endowed with other senses in place of these we now possess, or if our present meagre channels of communication with the outer world were opened wider, this world would seem very different to us. We can at best, so the argument runs, view the world through coloured spectacles which shut off all light except of those colours to which our senses are attuned —beings which could experience the full light of day would give a very different account.

Laboratory Data

Science has of course provided us with methods of extending our senses both in respect of quality and quantity. We can only see one octave of light, but it is easy to imagine light-vibrations some thirty octaves deeper than any our eyes can see. While philosophy is reflecting how different the world would appear to beings with eyes which could see these vibrations, science sets to work to devise such eyes—they are our ordinary wireless sets. We also have means for studying vibrations far above any our eyes can see. Actually a range of vibrations extending over about 63 octaves can be detected and has been explored—63 times the range of the unaided eye. And even this limit is not one of the resources of science, but of what nature provides for us to see. In the same way, the spectroscope makes good the deficiency of our eyes for analysing a beam of light into its constituent colours, and further enables us to measure the wave-length of each colour of light to a high degree of accuracy.

Science has extended the range and amplified the powers of our other senses in similar ways, in quality as well as in quantity. We cannot touch the sun to feel how hot it is, but our thermocouples estimate its temperature for us with great accuracy. We cannot taste or smell the sun, but our spectroscopes do both for us—or at any rate give us a better acquaintance with the substance of the sun than any amount of smelling or tasting could do. We are entirely wanting in an electric sense, but our galvanometers and electroscopes make good the deficiency.

Nevertheless, no one would claim to be able to imagine all the kinds of senses that we might possibly possess, or maintain that science has provided us with substitutes for them all. We can imagine beings who could neither see, hear, smell, taste nor touch, and yet were endowed with other senses, of kinds not only unknown to us but totally unimagined by us. Would their world be at all like ours?

The reply is that the instruments of research provided by physical science disclose a fairly self-contained region of phenomena. We may properly suppose that a reality, which we may describe as the physical universe, underlies it. Whether this is the whole of reality is a matter for debate.

Some biologists, for instance, believe that this domain includes the whole domain of biology; others prefer to think that a connecting passage-way leads from this domain of physics to a whole new domain of life. Again, those of a purely materialistic outlook maintain that the domain of physics comprises the whole of reality, while those who believe in the reality of a world of the spirit—the poet, the artist, the mystic—are at one in believing that there are other domains than that of physics.

In their more irrational moments, these latter may feel inclined to maintain that these other domains are entirely distinct from that of physics; that there are no connecting ways between them, so that the universe is not one but many. Such a contention can hardly survive serious reflection. The artist may often claim that his creations are "on a higher plane" than the purely physical, but he can hardly claim they are totally disconnected from it; no one knows better than he how much his imaginings depend on the state of his physical health and the condition of his physical tools and instruments. No poet will write quite the same "Ode to Joy" when he has a cold in his head as when he has not. And the preacher who has just told us how hard it is for the rich man to enter the Kingdom of God must not, at any rate in the same sermon, tell us that worldly riches are on a different plane from, and entirely unconnected with, the Kingdom of God.

For the moment let us merely remark that physical science is competent to discuss these questions. If passage-ways connect the domain of physics with the domains of life or of spirit, physics ought in time to discover these passage-ways, for they start from her own territory. When physicists are urged to investigate the claims of psychical science to produce ectoplasm, to speak with a "direct voice", to agitate tables and produce other material phenomena by non-material means, they are in effect being invited to decide as to the reality or not of alleged channels of communication of precisely this kind.

The Study of Nature

It will be convenient to conclude the present chapter by reviewing, very briefly, the history of man's efforts to understand the workings of the external world. We may distinguish three broad epochs, the nature of which may be suggested by the words animistic, mechanical and mathematical.

The animistic period was characterised by the error of supposing that the course of nature was governed by the whims and passions of living beings more or less like man himself. Before our infant can distinguish between animate and inanimate objects, he is destined to pass through a stage of confusing the two. He will fail to catch the rolling marble just as he failed to catch the crawling fly, and will assign the same reason—like the fly the marble was anxious not to be caught. He will trap his finger in the door, and attribute his sorrows to the naughtiness of the door. Because personality is the concept of which he has most immediate and direct experience, he begins by personifying everything.

As the history of the individual is merely the history of the race writ small, our race did much the same in its infancy as its individuals still do in theirs. Sometimes they endowed the inanimate objects of nature with wills of their own, sometimes they supposed them governed by the caprices of gods, goddesses, wood-nymphs and demons. A storm at sea was not the result of a depression moving eastward from the Atlantic, but of Poseidon and Boreas playing schoolboy jokes on their fellow-Immortals, or possibly even interfering in human affairs. As Menelaos is dragging the naturally reluctant Paris to slaughter by his helmet, the chin-strap gives way—not because its tensile strength was unequal to the pull of the indignant husband, but because (or so Homer tells us) Aphrodite herself loosens the strap as a return for favours previously conferred on her in the shape of a golden apple. This anthropomorphic fallacy permeated man's whole view of nature, as it still does in primitive races until scientific knowledge supersedes it. Such views of nature were un-reflecting and almost instinctive, arising in part from man's projecting his own personality on to nature, with a resultant confusion between man and nature, and in part from a mere fixation of infantile ideas.

Then in Ionian Greece, six centuries before Christ, the human intelligence began consciously to apply itself to the study of nature. It felt very little desire to increase its factual knowledge of nature, so that Greek science consisted in the main of mere vague questionings and speculations as to why things came to be as they were rather than otherwise.

It was not until the time of Galileo that science turned from cosmology to mechanics, and from speculation to experiment. The simplest way of affecting inanimate matter was to push it or pull it by means of muscular effort. So long as men could only experiment with objects which were comparable in size with their own bodies, they found inanimate nature behaving as though its constituent pieces exerted pushes and pulls on one another, like those we exert on them by the action of our muscles. In this way the science of mechanics came into being. Pieces of matter were supposed to exert "forces" on one another, and these forces were the causes of the motions of the bodies in question, or rather, the changes in their motions. And it was found that the behaviour of every object was determined, entirely and completely, by the pushes and pulls to which it was subjected; there was no longer any room for the intervention either of gods or of demons. A chin-strap broke as soon as the pull on it exceeded its tensile strength; no number of golden apples given to Aphrodite could have made it break sooner or later. The wind became high and the sea rough as soon as the barometric gradient exceeded a certain intensity—and so on. Bodies moved just as they were pushed or pulled by other bodies; nothing else mattered.

Science, having established these laws for objects of tangible size, went on to imagine that they governed the whole of nature. Thus when Newton (1687) had explained the motions of comets by mechanical concepts, he expressed a wish that the whole of nature might in time be explained on similar lines. Three years later, Huygens described the principle which was to guide physical science—wrongly as we now know—for the next two centuries, in the words: *

"In true philosophy, the causes of all natural pheno-mena are conceived in mechanical terms. We must do this, in my opinion, or else give up all hope of ever under-standing anything in physics".

* *Traité de la Lumière* (Leyden, 1690), Chapter 1.

Closely connected with this view of the workings of nature was the principle described as "The uniformity of nature". This asserted that, when the same experiment was performed any number of times on exactly similar objects in exactly similar circumstances, the result was necessarily always the same. The simple explanation was of course that the bodies under observation were subjected to the same pushes and pulls on the various occasions, and so behaved in the same way. Science admitted no exceptions to this uniformity; alleged violations of it were adjudged to be miracles, frauds or self-deceptions according to circumstances and the mentality of the judge. And just because observation and everyday experience seemed to establish this principle so firmly, scientists were wholly convinced that their simple mechanical explanation of it was the true one—bodies moved just as they were pushed or pulled by other bodies.

Causality, Determinism and Free-Will

This view of nature was soon seen to have far-reaching implications, and to raise far-reaching questions. When one object pushed or pulled another, the conditions prevailing at the moment determined the intensity of the push or pull. Science had now discovered that the intensity of this push or pull, and this alone, determines the ensuing motion of the affected object, which in turn determines the conditions that will prevail at the next moment, and so on. Thus conditions at one moment determine those at the next, these determine the conditions at the succeeding moment, and so on *ad infinitum*. The universe appears as a mere machine, wound up to go like a machine and destined to run down like a machine. Its whole future is inherent in its state at any moment, just as this state must have been inherent in its state at its creation.

This of course supposes that there is no intervention from outside, as, for instance, by the will-power of living things. Is it, however, conceivable that will-power should intervene? If the closed system of nature provides no opening for the activities of gods and goddesses, is it likely to leave a loophole for the similar activities of animals and men? The physiologists tell us that the brain is part of, and continuous with, the body; apart from reflexes, the atoms of our brains direct the motions of our bodies. On the mechanical view of nature, these atoms must move precisely as they are pushed and pulled about—like the atoms of a motor car. The butcher kills a lamb, and immediately the brain which had just been directing the creature's leaps and bounds, becomes mere mechanical matter—sheep's brains, to be thrown on the scales and sold by the pound. The metamorphosis has been accomplished without the loss or gain of a single atom. Why, then, should the quality of the pushes and pulls on the atoms of the brain change so abruptly just at the moment when the mind leaves the body?

The obvious suggestion is that these atoms experience pushes and pulls from mental as well as from material sources—in brief that our volitions can affect the atoms of our brains, and through them the atoms of our bodies. The plain man accepts this without even pausing to consider any alternative. He is quite sure that his mind is, within limits, free to guide his body, so that he can keep his appointments, and put a **X** where he pleases on his ballot-paper. He beats his dog for not coming when he whistles for it, and sends the forger to gaol because his fingers have written someone else's name at the foot of a cheque. All this provides him with a self-consistent scheme which is agreeable not only to his intuitions, but also to his moral sense.

The challenge to this scheme did not come, in the first place, from science but from philosophy; it originated with Descartes (1596–1650). His philosophy regarded mind and matter as entirely independent "substances", each existing in its own right apart from the other, and of such essentially different natures that they could not possibly interact—the one, for instance, existed in space, the other out of space. He accordingly thought of mind and matter as moving, so to speak, on parallel yet entirely different sets of rails, completely without interaction, and yet synchronised after the manner of a cinematograph film and its "talkie" accompaniment, so that the appropriate mental thoughts, moods and emotions always come at the right moment to suit the prevailing arrangement of atoms and the associated events.

A child seeing a speaking film for the first time might well think the words were the natural outcome of the events occurring before its eyes; it would be hard to believe that words which seemed to fit the events so naturally had been planned to match long in advance. So it is with our thoughts and the atoms of our world; they not only seem to match, but to emanate one from the other. Descartes, however, insisted, as in a different way did Leibnitz at a later date, that at the first morning of creation, a supremely benevolent God had miraculously arranged for a perfect and continuous synchronisation between bodily and mental events. Faith cannot really move mountains, because one is imponderable and the other so very ponderable, but the good God lets us think it can.

Descartes accordingly compared the body to "an earthly machine" actuated by a sort of reflex mechanism:

"You may have seen in the grottoes and fountains which are in our royal gardens, that the force with which the water moves when issuing from its source is of itself enough to set various machines in motion, and to make various instruments play or utter words, according to the different arrangements of the tubes which convey the water. We may compare the nerves of the machine which I am describing with the tubes of the machines of these fountains, the muscles and tendons with the other engines and springs which move the machines, and the animal spirits, the source of which is the heart and of which the cavities of the brain are the reservoirs, with the water which sets them in motion. Moreover, breathing and similar acts, which are natural and usual to the machine, and depend on the flow of the spirits, are like the movements of a water-clock or mill, which the ordinary flow of water keeps continually in motion. External objects, which by their presence act on the sense-organs of the machine and so determine it to move in different ways according to the disposition of the parts of the brain, are like strangers who enter one of the grottoes and are themselves the unwitting cause of the movements they witness. For on entering they tread on certain tiles or plates which are so arranged that if they approach a bathing Diana they cause her to hide in the rose bushes, and if they try to follow her they cause a Neptune to come towards them threatening them with his trident. Or if they pass in another direction they make a sea-monster spring forward and spout water in their faces, or things of a like kind according to the caprice of the engineers who constructed them".

He further, and quite inconsistently, compared the mind to an engineer in a control tower, who, by manipulating taps, could change the course of the water from one pipe to another with a minimum of effort, and went dangerously near to repudiating his own philosophy when he conjectured that the whole mechanism of our bodies was worked by "animal spirits" which were, he said, like "a very subtle air", the subtlety being so marked that they were on the verge of ceasing to be material. Later and more scientific writers have remarked that even in a completely mechanical world, large amounts of energy can have their courses changed by the expenditure of a minimum of energy, as, for instance, in moving railway points or turning over an electric switch.

These and many other elaborate fabrications arose out of a natural desire to avoid the implications of a mechanical view of nature. So long as his science appeared to tell him that his intuitive beliefs were erroneous, and the actions he based on them irrational, even the scientist himself could not but feel that his studies were divorced from reality, or at best were concerned only with a small corner of reality which had but little connection with his everyday life. Bradley, writing in 1899, summed up the current feeling in the words: *

"Nature to the common man is not the Nature of the physicist; and the physicist himself, outside his science, still habitually views the world as what he must believe it cannot be".

* *Appearance and Reality*, p. 262.

And, even as late as 1926, we find Whitehead writing:*

"The Western people exhibit on a colossal scale a peculiarity which is popularly supposed to be more especially characteristic of the Chinese. Surprise is often expressed that a Chinaman can be of two religions, a Confucian for some occasions and a Buddhist for other occasions. Whether this is true of China I do not know; nor do I know whether, if true, these two attitudes are really inconsistent. But there can be no doubt that an analogous fact is true of the West, and that the two attitudes involved are inconsistent. A scientific realism, based on mechanism, is conjoined with an unwavering belief in the world of men and of the higher animals as being composed of self-determining organisms. This radical inconsistency at the basis of modern thought accounts for much that is half-hearted and wavering in our civilisation. It would be going too far to say that it distracts thought. It enfeebles it, by reason of the inconsistency lurking in the background. After all, the men of the Middle Ages were in pursuit of an excellency of which we have nearly forgotten the existence. They set before themselves the ideal of the attainment of a harmony of the understanding. We are content with superficial orderings from diverse arbitrary starting-points. For instance, the enterprises produced by the individualistic energy of the European peoples presupposes physical actions directed to final causes. But the science which is employed in their development is based on a philosophy which asserts that physical causation is supreme, and which disjoins the physical cause from the final end. It is not popular to dwell in the absolute contradiction here involved. It is the fact, however you gloze it over with phrases".

Clearly a science which involved such implications and entailed such consequences was in need of a new background, such as should reconcile the nature of the

* *Science and the Modern World,* p. 94.

laboratory and the text-books with the nature of everyday experience. Happily it has acquired such a new background within recent years.

The New Physics

Throughout the mechanical age of science, scientists had proceeded on the same general lines as the child and the unreflective savage. Out of the impressions registered through their senses, they had built an inferential world of objects which they believed to be real, and affected by events of much the same kind as occurred in their everyday experience. They described this as the "common-sense" view of science; and defined science as "organised common-sense". Any scientific theory which could not be explained in terms of the familiar concepts of everyday life was said to be contrary to common-sense, and could hope for but a cold and unsympathetic reception, either from laymen or scientists. Then new refinements of experimental technique brought new observational knowledge, which shewed that the workings of nature could not be explained in terms of the familiar concepts of everyday life. New and unfamiliar concepts were found to be necessary; the age of common-sense science had passed.

Had science continued to pursue its old methods, it might have tried to draw concrete pictures of these new concepts. Some scientists indeed tried, by introducing a multitude of small changes here and there, to modify the old view of nature so that it could meet the new demands upon it. But they were trying to confine new wine in old bottles; their efforts met with no success, and the main stream of scientific thought followed a very different course.

For it was just about this time that science, mainly under the guidance of Poincaré, Einstein and Heisenberg, came to recognise that its primary, and possibly its only proper, objects of study were the sensations that the objects of the external universe produced in our minds; before we could study objective nature, we must study the relation between nature and ourselves. The new policy was not adopted of set purpose or choice, but rather by a process of exhaustion. Those who did not adopt it were simply left behind, and the torch of knowledge was carried onward by those who did.

This new line of advance has led us to a science which is no longer in flat contradiction with our intuitions and the experiences of everyday life; the physicist need no longer feel that his laboratory door divides his life into two watertight compartments—as scientist and as human being. In particular, mechanism, with its implications, has dropped out of the scheme of science. The mechanical universe in which objects push one another about like players in a football scrimmage has proved to be as illusory as the earlier animistic universe in which gods and goddesses pushed objects about to gratify their own caprices and whims. We are beginning to see that man had freed himself from the anthropomorphic error of imagining that the workings of nature could be compared to those of his own whims and caprices, only to fall headlong into the second anthropomorphic error of imagining that they could be compared to the workings of his own muscles and sinews. Nature no more models her behaviour on the muscles and sinews of our bodies than on the desires and caprices of our minds.

Whether determinism has also been banished from nature is still a question for debate. We shall see later that the answer is probably something more subtle than a mere "Yes" or "No"; possibly we could make either answer true by suitable definitions of determinism and nature. But that those particular causes which seemed until recently to compel determinism have gone—this is hardly open to question.

We shall see the fundamental contrast between the old science and the new very clearly if we compare the beginning of Newton's *Principia*, in which the mechanistic view of nature was first put in perfect logical form, with the beginning of Dirac's *Quantum Mechanics*, which represents the most complete exposition of the new theory of Quanta at present in existence.

Newton wrote in 1687:

"Every body perseveres in its state of rest, or of uniform motion in a right line, unless it is compelled to change that state by forces impressed thereon. The alteration of motion is ever proportional to the motive force impressed...";

and Dirac in 1930:

"When an observation is made on any atomic system that has been prepared in a given way and is thus in a given state, the result will not in general be determinate, i.e. if the experiment is repeated several times under identical conditions several different results may be obtained. If the experiment is repeated a large number of times it will be found that each particular result will be obtained a definite fraction of the total number of times, so that one can say there is a definite probability of its being obtained any time the experiment is performed".

CHAPTER II

THE METHODS OF SCIENCE

We have already noticed the inadequacy of the definition which describes science as organised common-sense. We ought perhaps rather to define it as organised knowledge. Such a definition makes it clear that the first stage in the development of any science must necessarily be the accumulation of facts. The facts may be either particular or universal. Some sciences, such as botany and pathology, still find it important to record exceptional and unusual occurrences which at first sight appear to form exceptions to the general scheme of nature. In the more exact and more highly developed sciences, such as physics and astronomy, there are none such to record; here nature appears to be governed by immutable laws. The aim of science is to discover and interpret these laws.

Scientific Synthesis

When a sufficient number of facts have been collected in any particular branch of science, the next stage is to try and cover them all by a general principle, which may or may not admit of an explanation in terms of familiar concepts. To be ultimately satisfactory, such a general principle or explanation must not only cover all the facts already known, but also all the facts which remain to be found out. It is accordingly first put forward in the form of a hypothesis. A scientist says in effect—"Observation shews that the following facts are true; I find that a certain hypothesis as to their origin is consistent with them all".

He and his colleagues may now set to work to obtain more accurate or extensive data bearing on the original facts, or entirely new facts may be discovered. The hypothesis may be tested by examining whether the extended and new facts can be covered, as the old were, by the proposed general principle or explanation. When two separate and conflicting hypotheses are in the field, it is sometimes possible to devise an *experimentum crucis* to decide between them. Suppose it can be shewn that if hypothesis A is true, a phenomenon X will occur, and that if hypothesis B is true, the phenomenon X will not occur. Then we can decide between the two hypotheses by performing an experiment, or taking an observation, to find whether the phenomenon X occurs or does not.

Interrogating Nature

Such an experiment, like every other, amounts in effect to asking a question of nature. This question can never be—"Is hypothesis A true?" but "Is hypothesis A tenable?" Nature may answer our question by shewing us a phenomenon which is inconsistent with our hypothesis or by shewing us a phenomenon which is not inconsistent with our hypothesis. She can never shew us a phenomenon which proves it; one phenomenon is enough to disprove a hypothesis, but a million million do not suffice to prove it. For this reason, the scientist can never claim to know anything for certain, except direct facts of observation. Beyond this, he can only proceed by building up hypotheses, each of which covers more phenomena than its predecessor, but each of which may have to give place to another hypothesis in due course. Strictly speaking, the time for replacing a hypothesis by a claim to certainty never arrives.

We have just considered the simplest possible instance of the process of interrogating nature. It is not always possible to frame a question which permits only of the answers "Yes" or "No". More difficult problems arise when the experimenter is deceived by hypotheses of his own imagining, and tries to obtain an answer to a nonsensical question. If his experiment can be carried through at all, it must give an answer of some kind, but the answer, when it comes, may well seem as nonsensical to the questioner as we can imagine the original question did to nature.

For instance, let us imagine a race of men equipped with perfect scientific instruments, but with very little scientific intelligence or knowledge. They see a rainbow in the sky, and wish to discover how far away it is. Treating it as though it were a piece of stage scenery, they instruct a party of surveyors to discover the distance of their cardboard rainbow. Observations taken with perfect instruments give a precise and unequivocal answer—the distance is *minus* 93,000,000 miles. Those in authority might decapitate their surveyors for incompetence, or harangue against the untrustworthiness of the observational method —"It is absurd to suppose that a distance can ever be negative, and anyhow 93,000,000 miles is preposterous, since the foot of the rainbow obviously lies between us and the mountain over there". But let them instead change the form of their question to nature, and express it in the form "How far in front of us is the source of the light we see in the rainbow?" and the answer, *minus* 93,000,000 miles, becomes full of significance. The preliminary *minus* now tells them that the source of light does not lie in front of them at all, but behind them, and as its distance is 93,000,000 miles they can at once identify it with the sun. It is frequently more difficult to frame a sensible question than to obtain an answer to a nonsensical one. And if the question was not rightly framed in the first place, it may be inconceivably difficult to interpret the answer aright.

To avoid the dullness and indefiniteness of a general discussion, let us step across forthwith to two particular instances, both of which will figure largely in the discussion that is to follow.

Astronomy and Relativity

The Greeks and Egyptians had collected a great array of facts concerning the apparent motions of the sun, moon and planets across the sky. About A.D. 150 Ptolemy of Alexandria attempted to cover them all by a single hypothesis. Contrary to the earlier views of Aristarchus of Samos and the Pythagoreans, he imagined the earth to form a fixed centre to the whole system, while sun, moon and planets revolved round it, the sun and moon revolving in circles, but the planets in a complicated system of cycles and epicycles. No new facts were brought forward to test this hypothesis, but in A.D. 1543 Copernicus brought forward an alternative hypothesis which appeared to explain the same facts in a simpler way; he supposed the sun, instead of the earth, to be the centre of the solar system, and earth, moon and planets to describe circles round it, the motions of the planets still being complicated to some extent by epicycles.

Two hypotheses were now in the field, and Copernicus devised an *experimentum crucis* to decide between them. If Ptolemy's hypothesis were correct, Venus could never appear as less than a half-circle of light. On the other hand, if Venus circled round the sun, its appearance, as seen from the earth, ought to shew phases like that of the moon, varying from a full circle down to a crescent as thin as that of the new moon. In 1609, the newly invented telescope provided the means of asking nature to decide between the two hypotheses. As soon as Galileo saw Venus appearing as a thin crescent of light, he knew that Ptolemy's hypothesis was untenable.

This did not of course establish the truth of the Copernican hypothesis. Indeed, new and more precise facts began to accumulate which threw doubt upon it. In particular, Kepler studied the motion of Mars in some detail, and found that this was inconsistent with the Copernican hypothesis. This led him to propound the new hypothesis that the planets did not move round the sun in cycles and epicycles, but in ellipses having the sun as common focus. For a time this hypothesis fitted all the facts known to astronomy.

Half a century later, Newton tried to combine these and other facts under the cover of a still wider hypothesis. He imagined every object in the universe to attract every other object with a force, the force of gravitation, which varied inversely as the square of the distance between the two objects, and supposed that the planets moved merely as these forces compelled them to move. He shewed that this hypothesis explained the elliptical orbits of the planets, and an immense range of other facts and phenomena as well—the motion of the moon round the earth, the fall of an apple to the ground, the parabolic trajectory of the cricket ball in flight, and even the ebb and flow of the tides. Finally it was found to account for the motion of comets. These fearsome and mysterious apparitions, which had hitherto been dreaded as portents of evil or symbols of divine displeasure, were now shewn to be mere chunks of inert matter, driven to describe paths round the sun by exactly the same forces as prescribed the orderly motions of the planets.

New data continued to accumulate, all of which fitted into Newton's theory, until in the middle of the nineteenth century the astronomer Leverrier found a discrepancy in the motion of the planet Mercury. Newton's hypothesis required a planet continually to repeat its path round the sun; it ought to describe the same ellipse again and again— like a small boy's engine running round and round the same track. Leverrier found that Mercury did not do this, but described an ellipse which itself turned round in space once every three million years or thereabouts. It was as though the track on which the toy locomotive ran was itself laid on a turntable, which slowly rotated in space while the locomotive ran rapidly round the track.

In time Einstein propounded yet another new hypothesis, the theory of relativity, which not only explained all the phenomena which Newton's theory of gravitation had previously explained, but also gave an accurate account of the motion of Mercury, and explained a great number of other scientific facts as well. It was possible to devise experiments and observations to provide crucial tests between the new theory of Einstein and the older theory of

Newton, and in every case nature ruled out the latter and decided in favour of the new theory. Other crucial experiments were designed to compare the new theory with the physical theories then prevailing, such as that light was propagated as waves in an all-pervading ether, and that electric and magnetic forces were transmitted as pressures and tensions through such an ether. Again, nature decided in every instance in favour of the theory of relativity. To-day, Einstein's theory provides an explanation of an enormous range of natural phenomena, and no single fact of nature is known to be inconsistent with it.

The general aim of science is to progress towards, and ultimately achieve, such theories. We can never say that any theory is final or corresponds to absolute truth, because at any moment new facts may be discovered and compel us to abandon it. Although this seems unlikely, facts as yet undiscovered may in time compel us to abandon the theory of relativity. But even if this occurs, the time spent in constructing it will not have been wasted; it will have provided us with a stepping-stone to a still wider theory, which will fit still more of the phenomena of nature. The layman sees Science, as it seems to him, for ever changing her mind, hesitating, turning back on her tracks, and repudiating her earlier opinions. The scientist sees her ever progressing through a succession of theories, each of which covers more phenomena than the predecessor it displaced, towards the goal of a single theory which shall embrace all the phenomena of nature. If such a theory is ever attained, it will give us a hypothetical scheme of the external world which will be capable of reproducing all the phenomena of the external world and no others.

Atomic Physics and Quantum Theory

Before turning to discuss the significance and value of such a scheme, let us take a second example of scientific progress, drawn this time from physical science.

When a mass of hydrogen gas is raised to incandescence —whether in the atmosphere of a hot star, or in an electric discharge in a terrestrial laboratory—the photons it emits prove to be of many different kinds, which can be specified by many different and distinct wave-lengths. A spectroscope will sort out the photons according to their wave-length, much as a potato-sieve sorts out potatoes according to their size, but with incomparably greater accuracy; the wave-length of the hydrogen photons can be measured to an accuracy of about one part in a hundred thousand.

There are reasons for thinking that each individual photon comes from a single hydrogen atom, which is believed to consist of one proton and one electron. For a long time it was difficult to see how so simple a structure could emit photons at all. The electron and proton were believed to be mere electrified particles which attracted one another according to the inverse square law. In this case, current theories of electric action shewed that the electron would describe an ellipse round the far more massive proton— just as a planet does round the sun—and would emit a continuous stream of radiation in so doing. There was therefore the primary objection that the emission of radiation would be gradual and not by complete photons. There was also the further objection that a gradual emission of energy would cause a gradual shrinkage in the size of the atom so that, contrary to observation, there could be no definiteness either in the size of the atom or in the quality of the photons it emitted.

In 1913, Dr Bohr of Copenhagen put forward a hypothesis which seemed for a time to dispose of all these difficulties. He supposed that hydrogen atoms could exist in a great number of different but quite distinct states, different amounts of energy being associated with each. There could be no gradual transitions between these states, but the atom might occasionally jump discontinuously from one to another, giving out energy in the form of a complete photon as it did so.

Some years later, Franck and Hertz of Göttingen obtained direct experimental evidence that such distinct states really existed. They found that when electrons collided with atoms, the latter might either take up certain large amounts of energy from the electrons, or none at all; they never took up a small amount of energy, so that a continuous dribble of energy was *a fortiori* impossible. The encounters were like a series of commercial transactions; money changed hands at each, but always by complete coins, so that each individual always had a certain number of complete coins in his pocket; fractions did not come into the question at all. And the amounts of energy associated with the different states were found to be precisely those required by Bohr's hypothesis.

Although this hypothesis was never quite consistent logically, it seemed to fit all the facts as known at the time. Then more refined measures of the wave-lengths of photons were obtained, and it was found that these did not completely agree with the predictions of the hypothesis. The hypothesis predicted the right results for the hydrogen atom under ordinary conditions, but the wrong results when the atom was put between the poles of a powerful magnet. It also gave wrong results for the normal helium atom, which is the simplest atom of all after hydrogen.

Recently a new hypothesis, forming what is known as the "new quantum theory", has removed at a single stroke both the logical difficulty and the whole of the observational discrepancies. The new theory is purely mathematical in form, dealing only with measurable quantities and the relations between them, but it admits of several physical interpretations. The best known of these, generally described as "wave-mechanics", supposes that electrons and protons are not mere particles of hard matter, as had previously been imagined, but that—much in the same way as photons—they possess many of the properties of waves.

Unlike Bohr's older hypothesis, this new hypothesis assigns to the atom properties which are in no way inconsistent with the inverse-square attraction of its electrons and protons; rather they are additive to it. The great merit of the new hypothesis is, however, that its predictions agree exactly with observation in every case in which comparison has been found possible. To consider the case of hydrogen light alone, it is probably an under-statement to say that twenty kinds of photons can have their wavelengths measured to one part in a hundred-thousand, and that in every case the measures agree, to within one part in a hundred thousand, with the values predicted by the new quantum theory. Now if this were a perfectly random hypothesis, having no relation at all to truth, only a piece of astounding good luck could enable it to predict even a single wave-length to an accuracy of one part in a hundred thousand; indeed there would be odds of something like a hundred thousand (10^5) to one against. The odds against the same luck holding good for a run of twenty wave-lengths would be something like 10^{100} to one against. Such at least would be the case if the wave-lengths were not inter-connected. Actually there is a certain inter-connection, since both the wave-lengths demanded by theory and those observed in practice fall into regular series. This circumstance obviously calls for a large reduction in the odds just mentioned, yet, even so, they remain enormously large—unthinkable millions to one against the agreement being a mere chance coincidence.

It should be added that the new quantum theory goes far beyond the explanation of the hydrogen-spectrum, or indeed of spectra of any kind; it explains a great number of phenomena in many departments of physics which had previously defied explanation, while not a single fact of observation is known to be inconsistent with it. Again, we see science moving towards a hypothesis which will cover all known facts with complete accuracy—if indeed it has not already attained such a goal.

The Search for Reality

Suppose, however, that two or more hypotheses prove equally well able to explain the whole range of phenomena. This is not a mere flight of fancy; in some restricted branches of science—the electromagnetic field equations, for instance—such a situation exists to-day. Is the scientist to rest content with two distinct and possibly inconsistent hypotheses, or shall he try to discover which of the two comes nearer to the realities of the external world?

The answer must of course depend to a large extent on what we regard as the ultimate aim of science. What is it that urges one set of scientists to spend arduous lives in discovering new facts to destroy old hypotheses, while another spends even more arduous lives in framing new hypotheses, destined to be destroyed in their turn by yet newer facts of observation? Up to now, the *raison d'être* of science has been irrelevant to our discussion.

Part of the value of science is of course utilitarian; it enriches our lives, and shews us how to live more comfortably and more happily—in brief, it lessens our pains and increases our pleasures. This is the obvious extension of the rudimentary science by which the one-day-old child tries to adjust itself to the hard facts of life.

Part of the value of science is intellectual. It would be a dull mind that could see the rich variety of natural phenomena without wondering how they are inter-related. Quite apart from all questions of practical utility, the modern mind feels strongly urged to synthesise the phenomena it observes, to try to combine happenings in the external world under general laws. This impels Karl Pearson to describe the function of science as "the classification of facts, the recognition of their sequence and relative significance". In the same spirit, Einstein writes: "The object of all science is to co-ordinate our experiences and bring them into a logical system". This view of the aims of science may take very extreme forms, as for instance when Dirac says that "the only object of theoretical physics is to calculate results that can be compared with experiment"—in other words to gratify intellectual curiosity, since otherwise it would be simpler to gain the required knowledge from the experiments direct.

These views regard science as being concerned solely with the phenomena of nature; the underlying reality from which the phenomena originate does not come into the question at all. And indeed many specifically maintain that the phenomena and their laws constitute the whole province of science—science, in brief, is concerned with what happens, not with what is. They hold that when science has included all phenomena in one single all-embracing hypothesis, she has run her course and nothing more remains for her to do. If two or more such hypotheses are in the field, well and good; either of them satisfies all requirements, and it is impossible to escape from the prison-house of the senses to discover which of the hypotheses agrees most closely with the external world. If we had a single picture which represented all the phenomena quite perfectly, we should have no means of investigating whether it represented reality or not.

Such considerations as this prove quite convincingly that we can have no certain knowledge of reality. They do not, however, touch the question of knowledge of probabilities.

For instance, we ask the question, "Can we know that the new quantum theory gives the true origin of the hydrogen spectrum?" The argument quoted above gives the answer, "No; we can know nothing of the external world", and a very satisfying answer it is—to one who does not wish to go any further. Science amplifies the answer, saying "No; we can know nothing of the external world *for certain*. At best we can only deal in probabilities. Yet the predictions of the new quantum theory agree so well with the observed spectrum of hydrogen that the odds in favour of the scheme having some correspondence with reality are enormous. Indeed, we may say the scheme is almost certain to be quantitatively true; that is to say, true to reality in those features which it is impossible to alter, in any way whatsoever, without destroying the numerical agreement of the theory with observation".

A probability which reaches so close a proximity to certainty is generally good enough for the ordinary affairs of life. It is far better than the kind of probability which lawyers describe as "good enough to hang a man on". Indeed, as Laplace remarked with reference to another scientific problem, it is far better than the probabilities in favour of the best attested events in history. We are accustomed to accept as an indisputable fact that Queen Anne is dead. The metaphysical argument that we can have no certain knowledge of anything beyond the confines of our prison-house, because we cannot go there to see, will of course prove that we cannot know this for certain. Indeed, it will take us as much further than this as we like; it can prove the impossibility of knowing that Queen Anne ever lived. But if we assume that she once lived, then no conceivable calculation can make the odds that she is dead anything like the odds of 10^{100} to 1, which we had occasion to mention just now. We may then argue that there is a better justification for supposing that the scheme we just discussed for the origin of the hydrogen spectrum is true in its numerical essentials than for supposing that Queen Anne is dead.

This particular argument only shews that we can acquire knowledge of numerical essentials—i.e. of factors which cannot be altered without destroying numerical agreement with observation. Other arguments might conceivably be devised to shew that other factors in reality can be known, at least to a high degree of probability.

Yet the possibility of acquiring knowledge of ultimate reality is obviously restricted by considerations which have already been mentioned. We cannot claim to have knowledge unless we can explain it to other beings with minds like our own. And we cannot explain, and so cannot know, the ultimate nature of external things except in the *a priori* improbable event of these proving to be of the same nature as something with which our knowing minds are familiar. For otherwise there is no standard of comparison, no language in which to describe it, for language can only describe experiences we have in common. Trying to explain reality, whether to ourselves or to one another, would be like trying to explain a wireless outfit to a savage. He would have no difficulty in understanding the phenomena, the voices or music that issue from the set, for he is accustomed to voices and music. He may even understand the atmospherics, for he is accustomed to thunder. Our troubles begin when we try to explain grid-bias, tuned circuits and high-tension batteries to him. And, except in the *a priori* improbable event just mentioned, we must expect to encounter similar difficulties when we try to explain reality either to ourselves or to others. It is this kind of difficulty, rather than the bleak metaphysical argument that we can have no certain knowledge of what lies beyond the confines of our prison-house, that constitutes the true barrier to progress.

Pictures of Nature

What we have so far described as a hypothesis might equally well be described as a picture, or a representation, or a model, of nature. It does not attempt to portray the reality of nature, but only what we see of nature—the phenomena of nature. It may reproduce all the phenomena within our cognisance with perfect fidelity, and yet may differ from reality in its essence just as much as a photographic print differs from the living face—colour, extension in a further dimension and all vital qualities may be lacking. The elements of this picture are necessarily concepts with which our minds are familiar, otherwise we could not have drawn the picture at all. On the other hand, the elements of reality need not be so, and if we cannot make our minds familiar with such elements as can exist in reality, we shall never understand reality. But it would seem that science might legitimately progress along the road from phenomena to reality by thinking over unfamiliar concepts until they become familiar, the concepts being selected in the first instance on grounds of probability, as appearing likely to figure in ultimate reality.

For instance, when the intelligent child is first told that the world is round, it at once protests that, if it were, the people on the far side would fall off; at a later age the concept of a round earth presents no difficulties. In the same spirit, the Victorian physicist used to say that he could never understand a physical concept of which he could not

have no existence in reality at all, but merely in our imperfect knowledge of reality.

This, however, brings us right up to the question which has been lurking in the background all the time—"What is reality?" I think it is possible that science and philosophy would answer this question in slightly different ways. The metaphysician is, I think, more inclined to regard reality and phenomena as detached and distinct, like a man and his image in a mirror, or an aeroplane and its shadow on the ground: to use a number of grotesque expressions, an entity may have either an ontal or phenomenal existence, but nothing in between. On the other hand, the scientist is more inclined to regard reality and phenomena as the two ends of a continuous road, along which it is his job to travel. The metaphysician may dismiss the statement that waves *really are* waves of probability as ignorant nonsense, while the scientist applauds it as a step towards final truth.

Other types of waves may not prove so easily intelligible as those we have just been discussing. Yet even here we may perhaps be able to discover certain properties which we may then try to visualise in terms of familiar concepts until finally our progress is stopped by something which we can neither picture, imagine nor describe. The Victorian physicist, for instance, used to picture light-waves as similar to the shakings of a jelly, or the waves of an earthquake, until he found that his picture did not agree with the facts of observation.

Again, we are familiar with the concepts of space, extension in space, limited extension in space, and so by a process of abstraction can pass to the concept of a particle. If our picture of nature proves to consist in part of particles, we again ask "Particles of what?" and may or may not be able to arrive at a partial answer.

Finally, we are familiar with the concept of mechanism through the interaction of our volitions and the muscles of our bodies.

It might conceivably have proved possible to picture the whole external world, completely and perfectly, in terms of familiar concepts such as waves, particles and mechanism; indeed nineteenth-century physics aimed consciously and deliberately at such a representation, not sufficiently realising how great the odds were against its being possible.

Had the attempt succeeded, science was all ready to identify the representation with the reality. Indeed, most scientists did this without waiting to see whether the representation could be made to fit all the facts of observation. It was usual to assert at this time that all discrepancies were sure to be cleared up in time, and those who taught science seldom allowed any other possibility to enter the mental field of vision of their pupils. Behind the scientists whole schools of philosophers, realists and materialists, were identifying reality with particles, waves and so forth out there in space. The few others who urged that neither the known facts nor any possible facts could compel or warrant any such identification were felt to be valiant defenders of a lost cause. Their voices passed almost unheeded, not because they could not prove their case, or because their opponents could prove a case against them, but because the probabilities at that time seemed overwhelmingly against them.

Our present observational knowledge shews that no

representation of this kind can fit the phenomena, so that the question of identification with reality does not arise. The external world has proved to be farther removed from the familiar concepts of everyday life than nineteenth-century science had anticipated, and we are now finding that every effort to portray it brings us up immediately against concepts which we can neither picture, imagine, nor describe. We have already seen that radiation cannot be adequately portrayed either as waves or as particles, or in terms of anything that we can imagine, and we shall soon find that the same is true also of matter.

Subjective Nature

The very real difficulties of modern physical science originate, in large degree, in the facts just cited. Physical science set out to study a world of matter and radiation, and finds that it cannot describe or picture the nature of either, even to itself. Photons, electrons and protons have become about as meaningless to the physicist as x, y, z are to a child on its first day of learning algebra. The most we hope for at the moment is to discover ways of manipulating x, y, z without knowing what they are, with the result that the advance of knowledge is at present reduced to what Einstein has described as extracting one incomprehensible from another incomprehensible.

Apart from this, science knows of only one way of proceeding so as to avoid a complete deadlock. Dividing the world up into (a) ourselves, (b) our experiments on the external world, and (c) the external world, it can leave off concerning itself with (c), and can concentrate on (b), our knowledge of the world as disclosed by experiments which we ourselves perform. The metaphysical argument mentioned above (p. 57) will suggest one obvious advantage of this procedure; it is that our knowledge of (c) can never consist of more than probabilities, whereas that of (b) will consist of certainties. But there is an even more immediate gain. However little we may be able to know the ultimate reality of external nature, and however unintelligible the imagined reality may be, the results of the experiments we perform on nature must necessarily be both knowable and expressible in terms of familiar concepts, since if the concepts had not previously been familiar, the experiments themselves would have made them so.

For instance, we experiment with light, and obtain results which are expressible in terms of the familiar concepts, waves and particles. The experiments do not tell us what the true nature of light is; they do not, for instance, tell us that it consists of waves, or of particles. They merely shew us light behaving in a way which reminds us sometimes of waves and sometimes of particles. We infer that the whole nature of light cannot be expressed by either of the words particles or waves, and as we do not know of any common object which is sometimes like waves and sometimes like particles, it may be that the true nature of light is for ever beyond our powers of imagining; quite certainly it is so now. Thus we cannot reason about light, only about the results of our experiments on light.

It is much the same with electrons and protons. We experiment with these (cf. frontispiece), and find that their behaviour reminds us sometimes of waves and sometimes of particles. As with light, one has yet imagined a consistent picture of what the electron and proton really are. At present the most we can do is to express quantitatively and in mathematical terms the properties of electrons and photons which our experiments reveal.

There is no compelling reason why this stage should be final, and it may possibly represent only a very transitory phase in the development of our knowledge. Our experiments on nature provide an obvious connecting bridge between ourselves and nature, and in exploring nature we naturally start from our own end of this bridge. Because the bridge involves ourselves as well as nature, it is hardly surprising that our present knowledge of nature should still possess a subjective tinge. For, after all, we only started on the right road a third of a century ago.

When we look into the future we see two possibilities. It may be that nature goes on her way regardless of us, and that it is only our imperfect present knowledge which involves ourselves as well. We can still only explore nature by stamping it with our own footprints and raising clouds of dust, so that our present pictures of nature shew our human stamp over it all. In time we shall perhaps learn how to remove our own footprints from the picture and shall then see that nature has a real existence, as much outside ourselves and independent of ourselves as the Sahara. The essentials of the Sahara are its particles of sand; the clouds we raise are transitory accidents. In 1899 most scientists would have unhesitatingly averred that nature was like this. Yet we shall see that up to the present science has hardly been able to find any solid ground behind the clouds.

There is another kind of desert in which cloud forms

are the essentials, and the medium in which they are expressed an accident—such are an artist's concept, or the traveller's recollections, of a desert. Nature may, too, be like this.

Broadly speaking, these two conflicting alternatives represent objectivist and subjectivist views of nature, or again realist and idealist schemes of philosophy.

Bradley wrote of the latter alternative: *

"It may be objected that we have now been brought into collision with common sense. The whole of nature, for common sense, *is*; and it is what it is, whether any finite being apprehends it or not. On our view, on the other hand,...the world of physical science is not something independent, but is a mere element in one total experience. And, apart from finite souls, this physical world, in the proper sense, does not exist. But, if so, we are led to ask, what becomes of natural science? Nature there is treated as a thing without soul and standing by its own strength. And we thus have been apparently forced into collision with something beyond criticism. But the collision is illusive, and exists only through misunderstanding".

Since this was written, science has gradually discovered that its nature "standing by its own strength" was an assumption rather than an ascertained fact, and so is more ready to admit that the collision may be illusive.

Yet the difficulties of the idealist position are almost too obvious to need description. A being who had no means of communicating with his fellow-men would have no means of knowing whether or not the nature he saw was a creation of his own mind; he might well credit it with no more real existence in its own right than the objects

* *Appearance and Reality*, pp. 279, 283.

he saw in a dream. We, on the contrary, must somehow fit into our scheme of nature the fact that, broadly speaking, innumerable other minds all observe the same nature as we do. Realism explains this very simply and naturally by supposing that nature exists outside of, and independently of, all our minds—we all see the same moon because the moon is out there, outside ourselves, for us all to see. Idealism cannot avail itself of this simple explanation; it has to suppose that our minds are in some way all members of one body, and so are all attuned to perceive the same concepts. They must be interconnected in some way—perhaps as the branches of a tree are interconnected, through having a common root—or perhaps again as the members of a shower of photons are interconnected; in some aspects these appear as a crowd of distinct individuals, in others as a continuous progression of light.

We leave the question here and proceed to discuss the findings of modern science, bearing in mind that they are a description, not of nature, but of human questionings of nature.

CHAPTER III

THE FRAMEWORK OF THE EXTERNAL
WORLD—SPACE AND TIME

We have already pictured the new-born child trying to correlate the events and objects which affect its senses, thereby taking its first steps towards becoming a scientist. Gradually it makes the discovery which we express by saying that the events can be arranged in time, while the objects in which they appear to originate can be arranged in space. Thus space and time form a sort of framework for the sense impressions which the child receives from the external world. The child does not of course concern itself with metaphysical questions as to the fundamental nature of space and time, and neither shall we here; only the simplest properties of space and time, as perceived by us, are relevant at the present stage of our discussion.

Rudimentary Views of Space and Time

The child finds that the events of its day come in simple sequence, like beads on a string. The string is what we call time, and the order of events relative to one another can be fully described by the words "earlier" and "later". Adjacent events need not be contiguous; just as there may be stretches of a string which are not occupied by beads, so the child may experience uneventful periods of time. Time passes through our minds like tape through a chronograph; any small fragment of it may or may not have events impressed on it. Somewhere in our physiological processes, a sort of clock ticks moments, and so gives us a sense of the passage of time. Through the tickings of this mental clock, our minds judge time-intervals to be long or short; we find that time passes through our minds in a way which is, approximately at least, the same for all of us, so that we are led to think of time as something outside ourselves, flowing past or through the consciousness of each of us as a river flows past the piers of a bridge. Science measures the flow of this supposed river of time more precisely by counting evenly spaced events—the passage of the sun or stars across the meridian, the ticking of a clock, the vibrations of a quartz crystal, or the oscillations of a tuned electric system. Until the theory of relativity compelled us to reconsider our position, we intuitively regarded time as an ever-rolling stream, whose flow could be measured in such ways as these.

Our intuitive conception of space is very different. Light enters our eyes from external objects, and our crystalline lenses arrange that all the photons (p. 28) which come from what we describe as "the same direction" shall be projected on to the same point of the retina. Our first classification of objects is accordingly by the points of the retina they affect, and, as the retina is a two-dimensional surface, we get the impression of objects arranged in a two-dimensional array of directions—angular space.

Yet we know that objects cannot be fully located by their directions alone. As we move about, they change their directions; sometimes a number may lie in the same direction, as seen by our eyes, and so interfere with one another's visibility. Looking in one single direction, I may see, one behind the other, tobacco smoke, a dirty window, a butterfly, a tree, a hilltop, a cloud, the sun. I arrange them in this particular order because of the way in which they interfere with one another's visibility. The arrangement is like that of events in time—a one-dimensional arrangement. The different directions of two-dimensional angular space each contain a one-dimensional arrangement of objects, so that objects, as they appear to me, form a three-dimensional array, which I can arrange in a three-dimensional "space".

Each of my two eyes makes such an arrangement independently for my mind, but something further is needed if both eyes are to make the same arrangement—as they must if the objects exist in their own right in the external world. We find that, just as consecutive events are not usually contiguous in time, so consecutive objects are not usually contiguous in space; the butterfly is not contiguous with my window, nor the cloud with the sun. Consecutive objects may be separated by "distance", just as consecutive events may be by time. Counting the ticks of a clock will give a measure of the time between events, and in the same way counting the number of end-to-end juxtapositions of a measuring-rod will give us a measure of the distance between objects. This particular way of measuring distance is, of course, independent of our sense of sight, and indeed of the properties and even of the existence of rays of light. Beings deprived of all senses but that of touch could still map out the arrangement of bodies in space, armed only with their sense of touch and a measuring-rod. Their arrangement might or might not agree with that of other beings who used only their eyes. It would agree if the straightness of rays of light was the same as that of the straight edge of a measuring-rod; otherwise not. The distinction is important because we shall see later that light does not always travel in such straight lines. Thus it is already clear that the arrangement of objects in space may have a subjective tinge about it; a blind man might make a different arrangement from one who could see and used no instrument except his seeing.

In some such way as this our individual consciousnesses first apprehend time and space. And once again the history of the individual is that of the race writ small.

Pre-relativity Views of Space and Time

We have seen how man as an individual only gradually becomes, and as a race only gradually became, aware of the existence of an objective nature, external to and independent of himself. What Professor Cornford describes as the "discovery of Nature...one of the greatest achievements of the human mind"* occurred in Ionian Greece six centuries before Christ. It is important—although, for the scientist, difficult—to realise that space and time were also human "discoveries" of about the same epoch. Jowett writes:†

"Our idea of space, like our other ideas, has a history. The Homeric poems contain no word for it; even the later Greek philosophy has not the Kantian notion of space,‡ but only the definite 'place' or 'the infinite'....When therefore we speak of the necessity of our ideas of space, we must remember that this is a necessity which has grown up

* *Before and after Socrates*, p. 15.
† *The Dialogues of Plato*, vol. IV, Introduction to Theatetus, p. 162.
‡ See p. 97 below.

with the growth of the human mind, and has been made by ourselves....

"Within or behind space there is another abstraction in many respects similar to it—time, the form of the inward, as space is the form of the outward. As we cannot think of outward objects of sense or of outward sensations without space, so neither can we think of a succession of sensations without time. It is the vacancy of thoughts or sensations, as space is the void of outward objects....Like space it has been realized gradually: in the Homeric poems, or even in the Hesiodic cosmogony, there is no more notion of time than of space".

Plato (*Timaeus*) describes space as*

"that which receives all bodies. It must be called ever self-same, for it never departs from its own quality....Were it like anything that enters into it, when things of opposite or wholly different character came to it and were received in it, it would reproduce them amiss, as its own features would shine through. Therefore also that which is to receive all kinds in itself must be bare of all forms, just as in the manufacture of fragrant ointments the artist first contrives the same initial advantage; he makes the fluids which are to receive his perfumes as scentless as he can. So, too, those who essay to model figures in some soft vehicle permit no figure whatsoever to be already visible there, but first level the surface and make it as smooth as they may....Space never perishes but provides an emplacement for all that is born; it is itself apprehended without sensation, by a sort of bastard inference, and so is hard to believe in. 'Tis with reference to it, in fact, that we dream with our eyes open when we say that all that is must be in some place and occupy some space, and that what is neither on earth nor yet in the heavens is nothing".

* Plato, *Timaeus*, Taylor's translation, pp. 49–51.

This view prevailed throughout the period of Greek science and until the time of Descartes (1596–1650); nature was conceived as consisting of solid objects interspaced with a characterless void, and the space of our intuition was regarded as a mere empty framework for the arrangement of these substantial objects.

Descartes introduced a new conception of space. It was fundamental to his philosophy that all substances fell into the non-overlapping and non-interacting categories of mind and matter, the essence of mind being thought, which did not occupy, and was not arranged in, space, while the essence of matter was occupancy of space and extension in space. He further maintained that all space must be occupied by something, arguing that empty space would fulfil no function, and that it was contrary to the perfection of design shewn throughout the universe that anything should exist without a purpose. Thus although the spaces between the stars might appear empty they could not be so, and must be occupied by some sort of continuous substance having a real existence and characteristic properties of its own. Space ceased to be a mere empty framework, and became an objective reality existing in its own right. This led Descartes to maintain that extension in space and motion through space were the true primary qualities of objects (p. 14).

In accordance with these ideas, Descartes abandoned the corpuscular theory of light, and imagined light to be of the nature of a pressure transmitted through this all-pervading substance to our eyes. At a later date, scientists also rejected the corpuscular theory of light in favour of the undulatory theory, which imagined light to be of the nature of waves. The all-pervading substance of Descartes could now perform the function of transmitting these waves. It was accorded a real existence, and described as the "luminiferous ether".

Location in Space

We have already noticed how two individuals might make different arrangements of the objects in space, according to whether they relied on their sense of sight or their sense of touch. It now appeared that nature, too, had her own special way of arranging objects in space, and this made all individual arrangements unimportant. They became right or wrong according as they agreed, or did not agree, with nature's own arrangement. Objects could not only be arranged in space; they could be located in space by their positions in the ether, just as objects in England can be located by their positions on English soil.

I can say for instance that an object is 50 yards north of the twentieth milestone on the Great North Road. If I tie my handkerchief to an object at this spot, take a walk, and come back to find my handkerchief still attached to the same object, I can say I have come back to the spot from which I started. On the other hand, if I drop my handkerchief overboard at sea, row about, and come back to my handkerchief, I am not entitled to say I have come back to the same spot, since currents and winds are likely to have moved my handkerchief. I can only fix a position at sea by taking bearings, directly or indirectly, from the land.

If space is occupied by an ether, we can locate a spot in space by the former method. We can, in imagination at least, tie a handkerchief to a particle of ether, and if we come back to the handkerchief we may say we have come back to the same point in space. We need not fear that currents and winds will have moved the handkerchief, for if light consists of waves travelling through an ether, its speed of travel shews that this ether must be far more rigid than steel.

If there is no ether, we can only locate a spot in space by its bearings from fixed landmarks, but where are such landmarks to be found? Not in the planets, for these are moving round the sun at speeds which range from 3 to 30 miles a second. Not in the sun and stars, which move past one another even more rapidly. Not in the great nebulae, the most distant objects known, for these are rushing away from us and from one another at still greater speeds of many thousands of miles a second. Nowhere in the whole of space can we find fixed landmarks from which to take our bearings, with the result that it is impossible to fix a position in space. Newton was fully alive to this difficulty, for he wrote:

"It is possible that in the remote regions of the fixed stars or perhaps far beyond them, there may be some body absolutely at rest, but impossible to know, from the positions of bodies to one another in our regions, whether any of these do not keep the same position to that remote body. It follows that absolute rest cannot be determined from the position of bodies in our regions".

He also saw how an all-pervading ether might provide a solution, for he continued:

"I have no regard in this place to a medium, if any such there is, that freely pervades the interstices between the parts of bodies".

And indeed the existence of such a medium would seem to provide the only solution of the problem; its particles would provide fixed standard positions, against which the positions of moving objects could be measured at any instant. If there is no such medium, we can only define rest in space in an arbitrary way.

Location in Time

The precise identification of instants of time presents problems of a similar kind.

We soon learn to regard the time of our own individual experience as an ever-rolling stream, and it used to be tacitly assumed that the same stream rolled on in the same way throughout the universe, so that events could be "located" in time, just as objects could be located in the ether. If, for instance, on January 1st, 1901, an astronomer saw a sudden outburst on a star which he believed to be 100 light-years' distant, he would say this outburst had occurred on January 1st, 1801. He believed that the outburst could be "located" in the stream of time, and that there was a definite meaning in saying that it had occurred at the precise instant of time at which the nineteenth century opened on earth.

Let us, however, consider what is implied in such a belief. It will be enough to consider a single illustration, taken from the every-day operations of practical astronomy. Let us suppose that British astronomers at Greenwich wish to compare their astronomical observations with those made by American astronomers at Annapolis, something more than 3000 miles to the west, and, with a view to doing this, set about synchronising their clocks. The obvious plan is to send some kind of a signal between the two places. If any known kind of signal travelled with literally infinite speed, the operation would be simple enough—the Annapolis astronomers would send out a signal when their clocks shewed noon, and if the Greenwich clocks shewed exact noon when this was received, the clocks would already be synchronous; if not, they could easily be adjusted to be so. The essential difficulties of the problem arise from the circumstance that no signal can travel with infinite speed, since it is a fundamental principle of physics that no signal can ever travel faster than light. Actually, astronomers use the fastest signals available, namely wireless signals, which travel at the speed of light. Yet if Annapolis sends out a wireless signal when their clocks shew noon, it will already be somewhat after noon at Annapolis by the time the signal reaches Greenwich. In practise the Greenwich astronomers say that as a wireless signal travels at about 186,000 miles a second, it takes approximately a fiftieth of a second to come from Annapolis. They therefore regard their clocks as adequately synchronised if they point to a fiftieth of a second after noon at the moment when the Annapolis signal reaches them.

This is near enough for the practical needs of astronomy, but it is not absolutely exact. To obtain perfect synchronism, it would be necessary to know the exact time which the signal took on its journey.

Now let us suppose that wireless signals consist of waves travelling through the ether, and imagine that the earth is also travelling through the ether, let us say in the direction from Greenwich to Annapolis. Then Greenwich would be advancing through the ether to meet the signal sent out from Annapolis, and so would meet it sooner than if the earth were standing at rest in the ether. But to know by how much sooner, and so discover the exact time of travel of the signal, it would be necessary to know the speed of the earth's motion through the ether.

The Michelson-Morley Experiment

The famous Michelson-Morley experiment tried to measure this speed in the most direct and most obvious way. If signals travelled through the ether at 186,000 miles a second, and the earth travelled through the ether from east to west at 1000 miles a second, signals travelling from west to east would have their rate of travel over the earth's surface increased from 186,000 to 187,000 miles a second because the earth would be moving to meet the signal, but that of a return signal from east to west would be decreased from 186,000 to 185,000 miles a second. A signal which made the double journey would be expedited on the outward journey, but retarded on the return journey. For each thousand miles of path, the outward journey takes $\frac{1}{187}$ second, the homeward journey $\frac{1}{185}$ second, so that we have as the total, per thousand miles of path:

$$\text{Outward time} = \tfrac{1}{187} \text{ sec.} = 0 \cdot 005347594 \text{ sec.}$$
$$\text{Return time} \;\; = \tfrac{1}{185} \text{ sec.} = 0 \cdot 005405406 \text{ sec.}$$
$$\text{Total time} = 0 \cdot 010753000 \text{ sec.}$$

On the other hand, if the earth were at rest in the ether, the total time would be:

$$\text{Total time} = \tfrac{2}{186} \text{ sec.} = 0 \cdot 010752690 \text{ sec.}$$

We see that the gain of time on the outward journey does

not quite make up for the delay on the return journey; there is a net delay of about a three-millionth part of a second.

Conversely, if the net delay could be measured, and proved to be a three-millionth part of a second, we should know that the speed of the earth's motion through the ether was 1000 miles a second.

Actually there was of course no means of comparing the time of the double journey with the time it would have taken had the earth been made to stand still. It was, however, possible to compare the times of two double journeys, both performed simultaneously on the moving earth, the one on the east-west course we have already considered, and the other on a course of equal length at right angles to this, and this comparison is found to give the needed information equally well.

To be precise, if we denote the speed of light by c, and the speed of the earth's motion through the ether by u, the loss of time per unit length of path on the double journey in the direction of the earth's motion is found to be

$$\frac{1}{c+u} + \frac{1}{c-u} - \frac{2}{c}.$$

Simple algebra shews that this is equal to

$$\frac{2}{c}\left[\frac{1}{1-\dfrac{u^2}{c^2}} - 1\right] \qquad \ldots\ldots(A).$$

Also simple geometry shews that the corresponding loss of time on the double journey in a direction perpendicular to the earth's motion is

$$\frac{2}{c}\left[\frac{1}{\sqrt{1-\dfrac{u^2}{c^2}}} - 1\right] \qquad \ldots\ldots(B).$$

The former quantity is easily seen to be very approximately double the latter, and if observation gives the difference between them, it is easy to deduce the value of u. In the actual experiment, the path of the rays was of course far less than the 1000 miles which we have taken for purposes of illustration; it was only a few yards, so that a speed of even a thousand miles a second would only have produced a time-difference of about a million-millionth part of a second. It is such minute times as this that have changed our whole outlook on the universe.

Michelson and Morley hoped to measure this small difference of time with accuracy, although naturally not with ordinary clocks or stop-watches. They performed their experiment with light of great purity of colour, the waves of which oscillate many millions of millions of times a second at a perfectly uniform rate, and these oscillations provided a very perfect clock. The experiment consisted in starting two beams of light simultaneously to run the out and home course in the two directions, and observing which got back to the starting-point first, and by how much it won.

If the difference of times had proved large, it would have shewn that the earth was moving rapidly through the ether; if small, that it was moving slowly. The one result that was never contemplated was that the time-difference should prove to be nothing at all. For the earth's motion round the sun alone gave it a speed of 19 miles a second, and the delicacy of the apparatus was such as to disclose a speed of only about one mile a second. Yet it was the unexpected that happened. When the experiments were performed, absolutely no time-difference could be detected. They have been repeated time after time, at different times of day and of the year (so as to get the apparatus pointing to different positions in space, and to get the earth at different parts of its orbit round the sun), under different conditions of temperature, altitude and so forth, but nature has consistently given the answer that she knows of no motion of the earth through the ether. The times given by formulae (A) and (B) are always precisely equal, so that $u = 0$.

At first such an answer seemed to be pure nonsense. The obvious inference (which, however, it took a very long time to reach) was that the question also had been non-sense—in brief, the concept of light as waves travelling through an ether had provided the wrong background for the experiment. The success of the undulatory theory shews that light has many wave-like properties, but these experiments seemed to shew that its mode of travel through space is not one of them.

The corpuscular theory had implied a different mode of travel. For if light travelled like waves through a sea of ether, its speed of travel would always be the same *relative to the sea of ether*. On the other hand, if it travelled like particles shot out from a gun, then its speed of travel would be always the same *relative to the gun from which it was fired*.

Perhaps then the question to nature ought to have been put in the form "Does light travel like waves or like particles?" When the question is framed in this way, the Michelson-Morley experiments unambiguously support the latter alternative.

Yet if light travelled like particles, the photons emitted by two bodies moving at different speeds would themselves move at different speeds. Now astronomical observation shews that the photons emitted by the two components of a binary star travel at precisely equal speeds, so that, in this case at least, light does not travel like particles. Clearly our last way of framing the question still assumed something we had no right to assume. It assumed that light must necessarily travel through space either as waves or as particles. Observation now seems to suggest that it does not travel as either.

How, then, does light travel through space? We shall see shortly how Einstein solved the puzzle by giving us a new conception, not of light but of space.

First, however, we must go somewhat back in the history of science. In 1873 Maxwell shewed that light was one special form of electric action, and the question of how light was propagated through space became only one aspect of a far wider problem. Both Maxwell and Faraday had tried to shew that all electric action was transmitted through space in the form of disturbances in the ether. Now it was obvious that, if the earth were travelling through the ether at 1000 miles a second, there would be what may be described as an "ether-wind" sweeping past and through all objects on the earth at a speed of 1000 miles a second. It seemed inconceivable that such a wind should not affect the transmission of electric action, yet experiments seemed to shew that it did not. A whole array of experiments on electric action in general gave information similar to that which the Michelson-Morley experiments had given about light. They not only failed to disclose the speed of the earth's motion through the ether, but seemed to indicate that no such motion existed. At any rate, the supposed ether-wind was found to have absolutely no effect on terrestrial phenomena.

Newtonian Relativity

Before discussing the significance of this, let us consider a simpler problem of the same nature, which had been discussed by Newton. It is well known that when a ship or train or other vehicle moves steadily forward at a uniform speed, objects inside it behave precisely as though it were at rest. If we play tennis on board ship, the player who is facing towards the bows of the ship gains no advantage from the ship's motion. Any advantage he gains in imparting speed to the ball is exactly neutralised by the extra effort needed to check its motion when it first impinges on his racquet. Actually it is a matter of common observation that the ball rebounds from our racquets exactly as though the ship were at rest. Newton expressed this fact of observation in the following words:

"The motions of bodies included in a given space are the same among themselves, whether that space is at rest, or moves uniformly forwards in a right line without any circular motion.

"A clear proof of which we have from the experiment of a ship; where all motions happen after the same manner, whether the ship is at rest, or is carried uniformly forwards in a right line".

and shewed why this must be in the following words:

"For the differences of the motions tending towards the same parts [i.e. in the same direction] and the sums of those that tend towards contrary parts, are, at first (by supposition), in both cases the same; and it is from those sums and differences that the collisions and impulses do arise with which the bodies mutually impinge one upon another. Wherefore (by Law 2) the effects of those collisions will be equal in both cases; and therefore the mutual motions of the bodies among themselves in the one case will remain equal to the mutual motions of the bodies among themselves in the other".

The same situation occurred when the action was electrical instead of mechanical; the motion of the earth was found to have no effect on the observed phenomena. Towards the end of the nineteenth century, a great number of physicists were engaged in investigating how this could be, and Professor Lorentz of Leyden announced a very remarkable conclusion in 1895.

The Lorentz Transformation

To make as vivid a picture as possible, let us imagine that a professor of physics discovered certain laws of electric action in a laboratory on earth, at some epoch when this happened to be standing still in the ether. Let us suppose that he formulated them in terms of measurements made in time and space. We may suppose he would follow the usual mathematical practice of specifying a point in space by its distances x, y, z from three perpendicular planes, and the passage of time by a quantity t which measures the interval which has elapsed since a specified zero hour. He can then express his law as a relation connecting certain quantities which admit of observation and measurement with x, y, z and t. If we want a concrete example to fix our thoughts, we may take the law of magnetic induction, which Maxwell expressed by the equation

$$\frac{1}{c}\frac{da}{dt} = \frac{dY}{dz} - \frac{dZ}{dy}.$$

Here c is the velocity of light; a is the magnetic induction in a certain direction, and Y, Z are electric forces in two other directions perpendicular to the first. Thus the law connects changes in the measurable quantities a, Y, Z with changes in x, y, z and t.

Now let us imagine that our physicist is subsequently shot out into space in a rocket which travels through space in the direction of x with a speed we may call u. He had discovered his laws in a laboratory through which no ether-wind blew, and so could hardly expect them to be true under his new conditions. Yet Lorentz was able to shew, from the known laws of electric action, that, notwithstanding the ether-wind, any laws of electric action which the physicist had discovered on earth would still be qualitatively true in the moving rocket. In a certain restricted sense they would also be quantitatively true. If he re-investigated these laws in the moving rocket, he would find that they could be expressed with perfect accuracy in precisely the same mathematical formulae as he had used on the earth at rest. The only point of difference would be that x, y, z and t would not have quite the same significance as they had on earth, although, as we shall shortly see, the physicist would never be able to discover this.

Let us reserve the symbols x, y, z, t for the measurement of space and time on earth; when the corresponding quantities are measured in the moving rocket, let us denote them by x', y', z', t'. Then Lorentz shewed that the same laws will be obtained in the rocket as on earth, provided the co-ordinates x', y', z', t' of the moving rocket are related to the co-ordinates x, y, z, t of the stationary earth by the equations

$$x' = \frac{x - ut}{\sqrt{1 - \frac{u^2}{c^2}}}, \quad y' = y, \quad z' = z, \quad t' = \frac{t - \frac{ux}{c^2}}{\sqrt{1 - \frac{u^2}{c^2}}}.$$

These equations express what is known as the "Lorentz transformation". Every term in them deserves careful study.

The symbol c still denotes the velocity of light as measured on earth; we shall soon see that it is also the velocity of light as measured in the moving rocket. When we are discussing problems of ordinary mechanics and astronomy, we need not trouble about the velocity of light at all; light moves so much faster than everything else, that we may quite properly think of it as travelling at infinite speed. Thus for the discussion of such problems, we may put c equal to infinity, as the mathematicians say, which means that everything divided by c becomes equal to zero. When we do this, the equations of the Lorentz transformation assume the much simpler form

$$x' = x - ut, \quad y' = y, \quad z' = z, \quad t' = t.$$

Thus, so far as mechanical experiments were concerned, our experimenter could use precisely the same co-ordinates in the rocket as he had previously used on earth, except for the difference $x - x' = ut$, which arises naturally from the circumstance that the rocket is increasing its distance from the earth at a speed u. This merely means that positions must be measured relative to the new laboratory, the rocket, and not relative to the old laboratory left behind on earth, as would naturally be done in any case.

Thus when the velocity of light is enormously greater than all the other velocities concerned, Lorentz's result becomes exactly identical with that which Newton had found more than two centuries earlier—all phenomena happen after the same manner, whether the laboratory is at rest, or moves uniformly forward.

Lorentz was, however, concerned primarily with electrical phenomena, which are known to be propagated with precisely the velocity of light, and so was not able to treat the velocity of light as infinite; this is why c appears in his formulae.

It first occurs in the factor $\sqrt{1 - \frac{u^2}{c^2}}$ in the denominator of x'. This means that x' is measured in different units from those in which the original x is measured. Just as twelve inches make a foot, so $\sqrt{1 - \frac{u^2}{c^2}}$ of the latter units make one of the former. This factor is sometimes described as the Fitzgerald-Lorentz contraction, because, while scientists still thought in terms of an ether pervading all space, Fitzgerald (1893) and Lorentz (1895) had independently suggested that an object which moved through the ether with a velocity u might undergo a contraction of this amount in the direction of its motion—the ether-wind might compress a body moving into it, just as the pressure of ordinary wind must compress a football kicked into it.

However we explain it, such a contraction is found to account exactly for the result of the Michelson-Morley experiment; the shortening of the apparatus in the up-and-down direction of the ether stream exactly compensates for the slower average speed of light on this course.

We shall see this at once if we turn back to the formulae (A) and (B) on p. 81. If the apparatus is shortened lengthwise by a factor κ when it is moving through the ether, the formula (A) giving the loss of time on an up-and-down journey of unit length must be replaced by

$$\frac{\kappa}{c+u} + \frac{\kappa}{c-u} - \frac{2}{c} \quad \text{or} \quad \frac{2}{c}\left[\frac{\kappa}{1-\dfrac{u^2}{c^2}} - 1\right],$$

and when κ has the value $\sqrt{1 - \dfrac{u^2}{c^2}}$, this becomes exactly identical with formula (B), which gives the loss of time on a crosswise journey.

No similar factor appears in the values of y' and z', so that there is no contraction in these directions; an object is only shortened in the direction of its motion, and not in directions at right angles to this. This leads to the odd result that motion alters the shape of an object; a billiard ball may be truly spherical when at rest, but ellipsoidal when in play. If Fitzgerald and Lorentz had been right, Gilbert's "elliptical billiard balls" would have described a sober scientific fact, provided that "elliptical" was meant to describe an ellipsoid of revolution. An object which moved with the speed of light would have been flattened to nothing at all in the direction of its motion; a sphere to a mere disc, a cube to a square, and so on.

The same shortening factor reappears in the value for

t', so that the experimenter in the moving rocket must measure his time also in units different from those he used in his laboratory on earth, if the motion of his laboratory is not to affect his description of the observed laws of nature. Again $\sqrt{1 - \dfrac{u^2}{c^2}}$ of the latter units will make one of the former.

The value of the time t' as measured in the rocket is not only complicated by the shortening factor in its denominator; the numerator also is complicated, depending not only on t, the time on earth, but also on x, the distance travelled from earth. This means that at any single moment on earth, when t has a known definite value, there is no corresponding definite value for t' which is the same at all points of space. For the man in the rocket, time varies at different points of space, just as "local time", or sun-time, varies at different points of the earth's surface. For this reason, Lorentz described the value of t' as the "local time" of the experimenter in the rocket. The astronomer's local time is propagated round the globe at such a rate that it is always "local" noon directly under the sun; Lorentz's formula shews that the physical experimenter's local time is propagated through space at a speed $\dfrac{c^2}{u}$.

Here we come upon a speed which is enormous even compared with the speed of light. If a rocket is moving at a ten-thousandth part of the speed of light, which is roughly the speed of the earth in its orbit round the sun, the "local time" for the rocket is propagated through space at ten-thousand times the speed of light. We shall see later that this speed of propagation plays a very important part in modern physics.

Let us take two concrete illustrations to explain the physical meaning of the Lorentz transformation. The law of electromagnetic induction given on p. 87 is expressed in terms of the co-ordinates x, y, z and t. Since we know the relation between these and x', y', z', t', it is merely a matter of algebra to express the law in terms of these latter co-ordinates. We find that it is expressed by the equation

$$\frac{1}{c}\frac{da'}{dt'} = \frac{dY'}{dz'} - \frac{dZ'}{dy'},$$

where a', Y', Z' have slightly different meanings from the original a, Y, Z. Physically this means that the experimenter in the moving rocket might re-investigate magnetic induction, and re-discover Maxwell's law. If he did, he could express it in precisely the same mathematical form as he had previously used on earth, although all the symbols except c would mean something a little different from what they had previously meant on earth.

As a second illustration, let us imagine that the experimenter in the rocket re-investigates the speed of propagation of light. Let us suppose that he ignites some magnesium powder at a point out in space which we may call the origin ($x = 0, y = 0, z = 0$), and at an instant which we may take to be zero-hour ($t = 0$). The flash of light produced in this way will set out to travel in all directions of space equally, at the same speed c. After a time-interval t, it will have travelled a distance ct in every direction, so that if it has reached the point x, y, z, whose distance from the origin is $\sqrt{x^2 + y^2 + z^2}$, we must have

$$\sqrt{x^2 + y^2 + z^2} = ct.$$

From this we can easily deduce, by using the equations of the Lorentz transformation, that

$$\sqrt{x'^2 + y'^2 + z'^2} = ct'.$$

Between them these two equations shew that whether the experimenter uses the modes of measurement appropriate to the moving rocket or those appropriate to the earth at rest, light will still appear to travel at the same uniform speed c in all directions. In other words, no number of Michelson-Morley experiments could possibly disclose the speed u with which the rocket was moving through space.

The Theory of Relativity

In 1905, Einstein gave a new and quite revolutionary turn to the whole problem. Lorentz had based his investigation on the concept of an ether filling all space, and consequently of an ether-wind blowing through every experiment. Consequently he had imagined that in some way the time t used by our observer at rest in space was real time, nature's own time, while the "local time" t' of the man flying through space in a rocket was merely a convenient fiction, introduced to allow for the ether-wind.

Yet if a new generation of men were born in the rocket as it moved through space, they would soon forget about the true time t they had left behind them on earth and would know of no time except the "local time" t'; this would be "the time" for them. In the same way, our human race knows only one time; we call it "the time", but actually it must be merely the "local time" of *our* rocket, the earth, as it moves through space. When we say that light from Sirius takes 8·65 years to reach us, we mean 8·65 years of the local time of our earth. When we say that an outburst on a certain distant star was synchronous with the beginning of the nineteenth century, we must be speaking in terms of the "local time" of our earth. It might seem obvious that we have no right to identify this with the "true time" of nature.

At this stage Einstein asked "Why not? What reason have we for supposing that our time is inferior to any other?" If the laws of nature are to be the same throughout space, all the various rockets moving through space with different speeds must have different local times, but there is no evidence that any true time exists which is superior to them all. Indeed, all the evidence points in precisely the opposite direction. True time implies the existence of a body at rest in space. Not only have we no means of discovering when a body is at rest in space, but there is every reason to suppose the phrase is meaningless.

On these grounds, Einstein maintained that all time is "local"; there are as many local times as there are rockets, or planets, or stars, moving through space, and none of them is more fundamental than any other.

This implies that it is just as impossible to locate an event in time in an objective way, as to locate an object in space in an objective way. Einstein accordingly proposed abandoning the concepts of objective, or absolute, time and space, and putting in their place the supposition, which all experimental evidence appeared to confirm, that "Nature is such that it is impossible to measure an absolute velocity by any means whatever". In brief, nature is concerned only with relative velocities; there is no fixed background of points in space against which motion can be measured in absolute terms, and consequently no absolute flow of time against which intervals of time can be measured.

The theory of relativity starts from this hypothesis, and proceeds to develop its logical consequences by strict mathematical analysis. If the hypothesis is true to nature, these consequences will agree exactly with the facts of nature. Many of them can be directly tested by experiment, and in every such case, without a single exception, nature has confirmed the theory—the consequences deduced from it have proved to be true. If ever one of these proved not to be true, it would at once become possible to measure an absolute velocity in space, and the observation in question would provide us with a framework of absolute space and absolute time. So far not a single physical experiment has done this, so that the picture which modern physical science draws of nature contains no reference to either absolute space or absolute time. We shall, however, see later that when astronomical science studies the universe as a whole, it may draw a slightly different picture.

This does not of course mean that we must abandon the intuitive concepts of space and time which we derive from individual experience. These may mean nothing to nature, but they still mean a good deal to us. Whatever conclusions the mathematicians may reach, it is certain that our newspapers, our historians and story-tellers will still place their truths and fictions in a framework of space and time; they will continue to say—this event happened at such an instant in the course of the ever-flowing stream of time, this other event at another instant lower down the stream and so on.

Such a scheme is perfectly satisfactory for any single individual, or for any group of individuals whose experiences keep them fairly close together in space and time—and, compared with the vast ranges of nature, all the inhabitants of the earth form such a group. The theory of relativity merely suggests that such a scheme is private to single individuals or to small colonies of individuals; it is a parochial method of measuring, and so is not suited for nature as a whole. It can represent all the facts and phenomena of nature, but only by attaching a subjective taint to them all; it does not represent nature so much as what the inhabitants of one rocket, or of one planet, or better still an individual pair of human eyes, see of nature. Nothing in our experiences or experiments justifies us in extending either this or any other parochial scheme to the whole of nature, on the supposition that it represents any sort of objective reality.

We used to think of space as something real and objective in the region "out there" from which messages came to our senses; it even seemed to acquire a sort of substantiality from the ether which we imagined to occupy its every point. We thought of time as something equally real and objective, flowing past our senses in a way entirely beyond our control. Yet when we question nature through our experiments, we find she knows nothing of either a space or of a time which are common to all men. When we interpret these experiments in the new light of the theory of relativity, we find that space means nothing apart from our perception of objects, and time means nothing apart from our experience of events. Space begins to appear merely as a fiction created by our own minds, an illegitimate extension to nature of a subjective concept which helps us to understand and describe the arrangement of objects as seen by us, while time appears as a second fiction serving a similar purpose for the arrangement of events which happen to us.

This is of course in striking contrast with the earlier views of Kant which had dominated metaphysics until the advent of the theory of relativity. These may be summarised as follows: *

"(1) The notion of Space cannot be derived from external experience; because, in order that I may apprehend things as out of me and out of each other, I must have the notion of Space already in my mind;

"(2) the notion of Space is a necessary, *a priori* one; for I cannot imagine Space annihilated, though I can very well think it emptied of objects".

In brief, for Kant, as also for Descartes and Newton, objects cannot exist without space; for Einstein, space cannot exist without objects.

Objective Space-time

We have seen our ordinary space and time becoming reduced to mere frameworks of human origin, against which we see and record our individual sense-experiences.

If we are to study objective nature, we clearly need an objective framework, which shall be independent of the motion of our particular rocket through space. Such a framework was all the time lying latent in the Lorentz transformation, although the genius of Einstein and Minkowski were needed to point it out. It is nothing more

* Sidgwick, *The Philosophy of Kant*, p. 38.

nor less than a four-dimensional space, having x, y, z and t for its four co-ordinates—in other words, the ordinary everyday space of any individual we please, extended by the addition of a fourth dimension, the ordinary time of the same individual. When the individual space and individual time of any particular individual are welded together in this way, the individual is found to drop out altogether—the constituents are subjective to a particular individual, but the product is objective.

An analogy from the ordinary three-dimensional space of everyday life will shew how this can be. We can divide ordinary space up in as many ways as we like; for many purposes it is found convenient to divide it into horizontal (two dimensions) and vertical (one dimension). Such a division is, of course, "local" to particular spots on the earth's surface; one man's vertical is not every man's vertical, and the division at London will not be the same as at Paris. Yet if an inhabitant of London combines his two-dimensional horizontal with his one-dimensional vertical, he will obtain just the same space of three dimensions as an inhabitant of Paris would obtain by the same procedure. Horizontal and vertical were local concepts, relative to London or Paris, but there is nothing local about the resulting space. Sometimes other modes of division may be more convenient than the horizontal-vertical division. An architect at work on the leaning tower of Pisa would probably use the division "perpendicular to the axis of the tower" and "along the axis of the tower". This would differ substantially from the horizontal-vertical division of the other inhabitants of Pisa, but would agree with the horizontal-vertical division of the inhabitants of Naples. The Pisan architect has a perfect right to use this division whenever he finds it convenient; it is not specially reserved for Neapolitans.

In the same way, each of us may divide our new four-dimensional space up into individual spaces and times in as many ways as we please. An individual often finds it convenient so to divide it that he regards himself as at rest; he thinks of the world as passing by him, rather than of himself as journeying through the world. At other times he may find other divisions more convenient. For instance, a terrestrial mathematician studying the motion of Jupiter's satellites would almost certainly choose a division which reduced Jupiter to rest in space—he would, so to speak, imagine himself living on Jupiter. Yet, however the division is made, when each man combines the space he has chosen with the corresponding time, the four-dimensional space he obtains will always be the same. The relation between one man's space and time and another man's space and time, or between the two spaces and times the same man may select for himself on different occasions, is of course given by the formulae of the Lorentz transformation.

Minkowski has shewn that this relation can be expressed in an even simpler form. If we write τ for ict, where i stands as usual for the square root of -1, and c is the velocity of light, the equations of the Lorentz transformation can be written in the form

$$x' = x \cos \theta - \tau \sin \theta; \quad y' = y,$$
$$\tau' = x \sin \theta + \tau \cos \theta; \quad z' = z,$$

where the angle θ is defined to be such that $\tan \theta = iu/c$. Every mathematician will see that these formulae represent a rotation of the axes of co-ordinates through an angle θ. To interpret them geometrically, we must think of a four-dimensional space in which x, y, z and τ figure as co-ordinates, just as x, y and z do in ordinary three-dimensional space; in fact, our new space is merely this ordinary space extended to a fourth dimension, having τ or ict for fourth co-ordinate. We now see that by turning the axes round so that they point in some new direction in this four-dimensional space—i.e. by rotating his individual directions of space and time in this four-dimensional space—one man may change his own space and time into those appropriate to another man, who is travelling through space at a different speed—just as, in ordinary space, by rotating his directions of horizontal and vertical through a certain angle, the Pisan may change his horizontal and vertical into those of the Neapolitan; his leaning tower has already rotated to shew him how.

The work of Lorentz, Einstein and Minkowski shewed in effect that although beings who are travelling at different speeds relative to one another will naturally divide up this four-dimensional space in all these different ways, they will all find the same laws of nature. In other words, nature herself has no special way of dividing it up. She is concerned only with the undivided four-dimensional space, in which she treats all directions equally. Such a space is generally described as a continuum. Clearly it forms the canvas on which we must draw our pictures of nature, if they are to be true pictures, free from all sub-jective bias. Indeed, we shall be able to test their truth by examining whether they treat all directions equally; as is said to be the case with modern cubist pictures, they must not suffer by being hung upside down or askew. Every picture or hypothesis which fails to satisfy this test must be discarded.

For instance, Newton's law of gravitation—that the force varies inversely as the square of the distance—fails to satisfy it. This is hardly surprising, since the "distance" between two objects has no precise meaning when we cannot synchronise time at the two objects. Coulomb's similar law of electric attraction also fails by itself, but magnetic forces step in to make good the deficiency, and electric and magnetic forces in conjunction are found to satisfy the test perfectly, as indeed we have already seen.

Objective Nature

Thus nature knows nothing of space and time separately, being concerned only with the four-dimensional continuum in which space and time are welded inseparably together into the product we may designate as "space-time". Our human spectacles divide this into space and time, and introduce a spurious differentiation between them, just as an astigmatic pair of spectacles divides the field of vision of a normal man into horizontal and vertical, and introduces a spurious differentiation between these directions. With astigmatic spectacles on, we incline our head and see the scene in front of us rearrange itself. Yet we know that nothing has happened to the objects in the scene. These are objective; our view of them through our spectacles is subjective.

When we take our human spectacles off, we see that an event no longer occurs at a point in space and at an instant of time, but rather exists at a point of the continuum, this point identifying both the time and place of its occurrence; we discover that the primary ingredients of nature are not objects existing in space and time, but events in the continuum. An object which was formerly characterised by continuity of existence in time may now be treated as a continuous succession of events—each event being the existence of the object at one instant of time, and one point of space. Thus an object is associated with a continuous succession of points, i.e. a line, in the continuum. This is commonly called the "world-line" of the object, its shape and position representing the motion of the object throughout its whole existence. Objects which are acted on by no forces, and so for ever move uniformly through space in straight lines, have straight lines in the continuum for their world-lines. If their speeds are the same, their world-lines are parallel; if different, they are inclined.

Two different events are of course represented at two different points, and the amount by which they are separated is known as their "interval". With our human spectacles on, we say that the interval between the departure of the Flying Scotsman from King's Cross and its arrival at Edinburgh consists of $7\frac{1}{2}$ hours in time, and 400 miles in space, but this is merely a private and subjective description—indeed, the fireman might describe it differently as $7\frac{1}{2}$ hours of hard work tied to a single spot—the footplate of the engine. When we take our human spectacles off, space and time fade away from view, and we see the departure of the train represented by a single point in the continuum, while another point represents its arrival. In the same way, with our human spectacles on, we say that the emission of a photon in Sirius and the reception of it by our eyes and instruments are separated by 51 million million miles, and by 8·65 years. When we take them off, we can only say that the two events are separated by so much interval in the continuum.

Certain philosophers object to this mode of treating the question, on the grounds that it presupposes that space and time have no existence in their own right, but only as seen by a conscious mind. They protest that the separation of the continuum into its two ingredients is physical and not psychological, so that, for instance, it does not require the mind, but only the body, of an observer—just as the selection of a rainbow out of the rays of the sun does not require the mind, but only the physical eye, of the observer; even a camera lens is adequate. We can test this contention by putting a dead body, say Imperial Caesar dead and turned to clay, into the continuum. We now have the continuum and the world-line of Caesar's body, and nothing else, and it is hard to find any sense in which "space-time" has been separated into space and time. We may of course agree to take the direction of the world-line at each point of it as the direction of time, and the other three directions as space, so that if Caesar returned to his body, he would not think of himself as travelling through the world but of the world as processing past him. Yet if we do this, the separation has not been effected by Caesar's dead body, but by our live minds. We cannot argue that Caesar's mind would necessarily effect the separation in this way, if he returned to life. When I am climbing a mountain, I do not choose my space and time in this way; I think of myself as going up the mountain and not of the mountain as coming down to me, and we need not doubt that Caesar used to do the same.

We must recollect that the space and time with which the theory of relativity deals admit of perfectly precise definition; they are the space and time which an observer, discovering or verifying or discussing laws of nature, chooses with his conscious mind as the framework against which to record his observations. The space and time of his choice may or may not coincide with the space and time of his conscious perception at the moment. The theory of relativity knows nothing of the latter, so that if we identify the two, it is at our own risk.

The space and time of relativity are definite and precise; often those of our conscious perception are not. When we voyage through a rough sea, the solid structure of the ship suggests one space visually to our consciousness, the horizon suggests another, while the combination of gravity and the every-varying accelerations of the ship suggests a rapid succession of quite different others, the continual conflict adding much to the woes of the unseasoned traveller. To take a more placid example of the same thing, while an astronomer is taking observations, a driving clock keeps his telescope pointing in a fixed direction in space, but his body shares in the earth's rotation. He will almost certainly choose to record his observations with reference to the fixed direction in space of the telescope, but unless he allows the space of his conscious perception to alternate repeatedly between this and the terrestrial space in which his body is at rest, he will find the telescope running away from his seat. When he jumps off a moving omnibus, he must change the space and time of his perception with

extreme alacrity or else he will fall. Yet if he subsequently
wishes to understand why he fell, he must choose either the
omnibus or the road as the framework for his calculations,
and must definitely confine himself to the one or the other:
he must in fact pass from the space-time of his perceptions
to that of the theory of relativity. Because the two are so
entirely different, the technique of avoiding a fall is the
exact opposite of that of understanding it after it has
occurred.

Past, Present and Future

Even when space and time are completely welded together
in the continuum, we can still distinguish two distinct
kinds of interval. It is a clearly established law of physics
that no material object can travel faster than a ray of light,
so that the speed of light—which we have already (p. 93)
seen to be objective, the same for all travellers in space—
provides an absolute maximum speed. If two events are so
located in the continuum that a body can be present at
both, although not travelling at a speed greater than that
of light, we say that the interval between them is "time-
like". Thus the interval between any two events on the
world-line of the same body—as for instance the departure
of the Flying Scotsman from King's Cross and its arrival
at Edinburgh—must always be a time-like interval. In
the same way, all the events which affect the individual
consciousness of any one of us are separated by time-like
intervals. It is from this we get our intuitive conception
of the flow of time.

On the other hand we say that two events are connected
by a "space-like" interval when a messenger would have

to move faster than light to be present at both. A boundary line between the two kinds of intervals is formed by cases in which a messenger could be present at both events by travelling with exactly the speed of light. For mathematical reasons, which do not concern us here, the interval in such a case is described as a "zero-interval". When events are separated from us by a time-like interval—e.g. the death of Queen Anne or the Coronation of King George—we can only know of them by the exercise of memory, or by the use of records which, by their permanence, arrest the flow of time. When events are separated from us by a space-like interval, we cannot know of them at all; more time must elapse until the interval becomes first zero, and then time-like, so that we can know the events. But when the interval is exactly zero, we can have direct and immediate knowledge of the events—the knowledge of seeing them with our own eyes.

So long as a river of time was supposed to flow equably through all points of space, events could be divided perfectly sharply into past, present and future. All the events of the world could be represented in a continuum constructed of three directions of objective space and one of objective time. A surface drawn through the three directions of space at any instant of time had the whole of the past on one side of it, and the whole of the future on the other. Itself, it contained the whole of the present.

The theory of relativity has shewn us that such a division is merely the private choice of a particular individual. The surface through the three directions of space of any individual still forms the "now" for that individual and divides his subjective time into his past, present and future.

It is sometimes suggested that by changing his speed through space, any man can wave his "now" about in the space-time continuum, much as the man in charge of a searchlight can wave his beam of light about in ordinary space; he can re-divide the continuum into past, present and future, much as the searchlight operator can re-divide space into darkness, light and darkness. Indeed, he need himself do nothing. If he sleeps for eight hours, the rotation of the earth will have changed his speed through space by several hundreds of miles an hour, and will have rearranged his division of the continuum accordingly. It may well have shifted ten years of time on a distant nebula from the past into the future, and so may seem to give its inhabitants ten years of their lives to re-live for good or for evil—or would it be merely to re-live precisely as they had already lived them before? The paradox disappears if we remember that the time involved is merely that which an individual chooses for the recording of his observations of nature. It is not the time of his consciousness, still less that of the consciousness of the inhabitants of the nebula. We cannot wave anything about in the continuum which is more tangible than our own thoughts.

Nevertheless, we see that time, as one dimension of the continuum, may be lacking in one of the properties with which our uncritical intuitions had endowed it, namely, its sharp division by a "now" into past and a future. We cannot prove either that such a division exists or does not; the theory of relativity merely suggests that when our intuitions suggested that it certainly did, they may have been misleading us.

Again, it has been suggested that if this sharp line disappears, the concept of evolution in time may lose all meaning. We used to think of the universe evolving much as a pattern is woven in a loom. Space and time were the warp and the woof of its weaving. At any one instant, so much and no more has been irrevocably fixed; the rest still lay hidden in the loom, the womb of time, to be brought forth in due course. On the mechanistic view of nature, the loom had been set to work according to certain unalterable laws, so that the complete pattern was potentially existent from the outset and evolution became a mere synonym for the disclosing of predetermined changes, the tapping out of a pattern already designed. On a non-mechanistic view the loom was guided no one knew how, and might produce no one knew what.

The theory of relativity in no way compels us to give up this simple picture of evolution, but it certainly casts doubt on the intuitive concepts on which it was based. And many have thought that, because of this, the whole comparison between the evolution of the world and the weaving of a pattern is faulty. Time, they say, no longer remains time or involves change; it is merely a geometrical direction of our own choice in the continuum. The pattern is not being continually woven piece after piece in a time which no longer exists, but is spread before us complete in a continuum in which future events have just the same kind of existence as past events. We say that Australia exists although we are not there to see—we have perhaps never been there yet, but shall go there some day. In the same way, they argue, may we not say that the year 1942 exists? We have not been there yet, but perhaps we shall get there some day. Indeed an inhabitant of the nebula we just mentioned can wave his "now" through the continuum until its intersection with the world-line of our earth passes instantaneously from 1932 to 1942, like the searchlight operator waving his beam of light over the clouds. The clouds are there whether the searchlight falls on them or not, and, so it is said, is the year 1942.

Again we must recall that the space and time with which the theory of relativity deals are merely a time and space selected by our own minds for the discussion of natural laws. The theory of relativity does not assert that anything more tangible than our own thoughts can impinge on the year 1942, and this can hardly be said to endow the year 1942 with a real existence at the present moment.

The theory of relativity is built upon a perfectly definite and concise experimental basis; in the analogy we have already used, it is that the whole of physical nature follows us about like a rainbow, or like our own shadow. The result is that we cannot find evidence of our own motion by questioning physical nature, so that absolute space and absolute time do not enter into the nature we study in our physical laboratories. But this is not to say that they cannot exist in a wider external world than that of pure physics. Indeed, we shall see later that astronomical nature finds some evidence of absolute space and absolute time, and this has a bearing upon the questions we have just discussed. We shall return to this later (p. 144).

From the time of Plato onwards, philosophic thought has repeatedly returned to the idea that temporal changes and the flux of events belong to the world of appearances only and do not form part of reality. The reality, it is thought, must be endowed with permanency, otherwise it would not be real, and we could have no knowledge of it—behind the kaleidoscopic changes of nature there must be a permanent kaleidoscope, imparting a unity to the flux of events.

For this kind of reason philosophers have insisted that reality must be timeless, and time merely, in Plato's phrase, "a moving image of eternity". Bradley,* for instance, writes:

"Change, as we saw, must be relative to a permanent. Doubtless here was a contradiction which we found was not soluble. But, for all that, the fact remains that change demands some permanence within which succession happens. I do not say that this demand is consistent, and, on the contrary, I wish to emphasize the point that it is not so. It is inconsistent, and yet it is none the less essential. And I urge that therefore change desires to pass beyond simple change. It seeks to become a change which is somehow consistent with permanence. Thus, in asserting itself, time tries to commit suicide as itself, to transcend its own character and to be taken up in what is higher".

And again, two pages later:

"Time is not real as such, and it proclaims its unreality by its inconsistent attempt to be an adjective of the timeless. It is an appearance which belongs to a higher character in which its special quality is merged. Its own temporal nature does not there cease wholly to exist but is thoroughly transmuted. It is counterbalanced and, as such, lost within an all-inclusive harmony.... It is there, but blended into a whole which we cannot realize".

* *Appearance and Reality*, pp. 207, 209.

We may notice how the absorption of space and time into a higher unity, the space-time continuum, which transcends both and is changeless, satisfies the requirements of the philosophers, although only at the expense of relegating evolution to the realm of appearance.

CHAPTER IV

MECHANISM

Action at a Distance

Primitive man saw nature as a collection of objects which acted on one another, if at all, by direct contact; he was familiar with the pressure of wind and water on his body, the fall of raindrops on his skin, the thrust by an enemy, but action at a distance was somewhat of a rarity in his scheme of things.

Early science hardly advanced on this view, picturing matter as consisting of hard objects, no two of which could occupy the same space because one invariably pushed the other out of the way by direct contact. The science of a later era, however, found many instances of action at a distance. A magnet attracts iron filings to itself from a distance, and is itself acted on by the yet more distant magnetic poles of the earth; two electrified bodies attract or repel one another across the intervening space according as they are charged with opposite or similar kinds of electricity; the sun attracts the planets, and the earth the falling apple. In none of these cases can anything tangible be found to transmit the attractions and repulsions. It is true that the space between the interacting objects will often be occupied by air, but this does not transmit the action; electrified bodies and magnets attract rather more forcibly in a perfect vacuum than in air, while an apple falls more freely and rapidly when there is no air-resistance to break its fall. The sun attracts the planets across a space which is practically void of matter.

At a still later period, matter was found to be wholly electrical in its structure, consisting of particles which carried electrical charges, and of nothing else. These particles were so minute that an object occupied enormously more space than the aggregate of the amounts occupied by its separate particles. Roughly a ton of bricks occupies a cubic yard, while the millions of particles which form this ton of bricks occupy only about a cubic inch; all the rest is empty space. The particles of the brick hold one another at arm's length through the electric forces they exert on one another. If these forces could be abolished, we could pack all the particles of a ton of bricks within a cubic inch of space. In the interiors of the densest stars the particles are packed as closely as this; the electric repulsions are not actually abolished, but they count for nothing against the immense forces resulting from the pressure of the star itself.

In ordinary everyday life, however, these electric forces maintain their supremacy against all others, and the pushes and pulls of common objects are as much the out-come of action at a distance as is the attraction of a magnet for iron filings or of the pole for the compass-needle. When the wind blows on my face, the molecules of air come to within about a thousand-millionth part of an inch of my skin, but no nearer; at this distance the molecules of my skin repel them so violently that they turn back the way they came. The sensation of the impact of the wind on my face is the outcome of the reaction of the electric forces exerted by the molecules of my own skin—just as I feel a reaction in my foot when I kick a football.

It is the same throughout nature. When we look at

it through a sufficiently powerful mental microscope, we find no instances of actual contact; nature appears to have only one mechanism, which is action at a distance—action across intervening space.

For a long time it was thought that the ether, which had been originally introduced to transmit waves of light, might transmit all these other actions as well, and so serve as the general mechanism underlying nature. Innumerable experiments were tried in the hope of discovering any signs of the existence of an ether. They one and all failed, which of course only amounts to saying that all the phenomena of nature were found to conform to the principle of relativity. Space appeared to be entirely empty except in the isolated regions which were occupied by objects.

How, then, was the action transmitted from the magnet to the iron filings, from the earth to the falling apple, from the moon to the tides, from the cricket bat to the ball? If the ether was no longer available for this purpose, something else must be found to take its place. The story of the quest for this new something brings us to the very heart of modern science.

The Curvature of Space

We have an intuitive belief that space is flat or Euclidean— parallel lines never meet, and so on. This is based on our everyday experience. Yet all that this actually tells us is that we can bring law and order into the arrangement of the objects with which our everyday life is concerned by imagining them arranged in a space of this kind. We might try other arrangements of objects, as for instance arrange-

ment in a four-dimensional space, and should soon discover that the ordinary Euclidian three-dimensional space,
which the layman describes as "space" without any ad-
jectives, had some sort of pre-eminence, at any rate for
the arrangement of such objects as we encounter in
ordinary life. Yet we have no right to assume that the
whole universe could be reduced to law and order by
being arranged in such a space; if we do so, we merely
repeat the old mistake of thinking that all nature is like
the small fragments of it with which we are familiar—the
"common sense" view of nature.

There is a certain peculiar sect whose members insist
that the earth's surface is flat, so that parallel lines drawn
on it can never meet. Their intuitive concept of the earth's
surface is like ours of space; both are based upon an
imperfect acquaintance with the whole. And, just because
the concept is intuitive, no amount of abstract argument
will persuade the man who holds it that it is faulty. The
little bit of the earth in which his daily walk or daily
labour lies is flat, and he absorbs this flatness into his
mental make up, until he is unable to conceive any possibility except flatness, which he then wrongly extends to
the whole earth. It becomes a matter of common sense
to him that the earth not only is, but must be, flat.

Suppose, however, that a member of this sect took to
travel. He might find it of interest to draw a map of the
earth on which to record his journeys—it would of course
be a flat map, as for instance an ordinary Mercator projection, which can be found in any school-atlas. When he
travelled by sea, he might copy down the ship's position
day by day, and in this way record the course of the ship

on his map. Now the shortest course between two points in the northern hemisphere always bends towards the north pole, so as to "take advantage of the shorter degrees of longitude", while in the southern hemisphere there is a corresponding deflection towards the south pole. For instance, the shortest course from Southampton to New York goes farther north than either Southampton or New York; the shortest course from Cape Town to Cape Horn goes farther south than either. When such courses are mapped on a Mercator projection they look very curved; a shortest course only looks straight if it happens to lie either due north and south, or else dead along the equator. No doubt our traveller would at first be surprised to find that all the steamers he patronised seemed to follow very curved tracks; he might imagine that they were pulled out of their direct courses by forces emanating from the two poles of the earth.

One day, however, he might shew the curved tracks on his map to a friend, and discuss their meaning with him. It would be a tremendous revelation if his friend took him to a spherical globe on which the countries of the earth were marked in their proper positions, stretched bits of string from point to point, and shewed him that when the string was pulled tight so as to give the shortest path from point to point, it invariably lay exactly over the course which had been followed by the steamers in his travels. He would then see that the ships had actually been following the shortest courses on a curved earth. Their tracks had appeared curved on his map, not because forces were pulling them out of their courses, but because the framework of latitude and longitude, which actually

is twisted by the curvature of the earth, had been arti-
ficially untwisted in his Mercator projection. In brief, he
had been trying to describe his journeys against a background which was not true to nature. He would discover
that, although a flat map was perfectly suited to the
arrangement of places in the immediate neighbourhood
of his own home, it was not at all suited to the representation of the whole of the earth's surface.

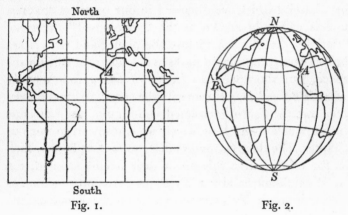

North

South

Fig. 1.   Fig. 2.

Notwithstanding its apparent curvature, the course from *A* to *B*
shewn in fig. 1 is the shortest possible. If it is mapped out on a
globe, as shewn in fig. 2, a tightly-stretched string will be found
to cover it exactly.

Newton's theory of gravitation had explained the curved
paths of planets, comets and cricket balls precisely as our
flat-earth traveller had explained the curvature of his
steamer tracks. The latter imagined that his steamers sailed
in a flat sea and were drawn out of their straight courses
by a pull emanating from the poles of the earth. Newton
imagined that the planets swam in a flat space and were

drawn out of their straight courses by a pull emanating from the sun; he imagined that the cricket ball was thrown in a flat space but that its course curved earthward because of a gravitational pull emanating from the earth.

We have already noticed that this theory of gravitation does not conform to the requirements of the theory of relativity. Einstein replaced it by a new theory which does; it is an extension of the simple or "restricted" theory of relativity which we discussed in our previous chapter, and is generally known as the "generalised" theory of relativity. It does not picture the planet and the cricket ball as describing curved paths in a straight (or Euclidean) space, but shortest paths in a curved space. Actually what is curved is not primarily the space of our ordinary life, but the four-dimensional continuum, the objective blend of space and time which we considered in our last chapter. The theory supposes that gravitating bodies, such as the earth, curve this up in their neighbourhood, the word neighbourhood now implying proximity in time as well as in space. This curvature deflects the planet and the cricket ball much as the molehill on the bowling-green deflects the bowl. Just as the friend of the flat-earth heretic was able to shew him a curved surface—in this case a sphere—in which all his complicated curves became shortest courses, so Einstein has shewn the scientific world a curved continuum in which the complicated tracks of planets, cricket balls, rays of light, and so forth, all reduce to shortest courses. As the continuum is curved, the space of our everyday life, which is a cross-section of it, must also be curved.

We notice that action at a distance has fallen out of the

picture. If we fix our attention on the three space-dimen-
sions of the continuum, we may say that the earth keeps
space in its neighbourhood continually curved, so that
when the cricket ball is thrown through this space it describes a curved path, much as though it were rolled along
a hill-side. The proximate cause of the curvature of its
path is the contiguous space, not the distant earth. True
action at a distance—action transmitted instantaneously
across intervening space—inevitably had to drop out, because of the impossibility of synchronising time at two
distant points.

The generalised theory is concerned with precisely the
same space and time as the restricted theory, namely, the
space and time which an investigator or scientist chooses
with his conscious mind for the recording of his observations of nature. He has a right, just as our traveller had,
to record these on any kind of map he pleases. When our
traveller tried a flat map he had to introduce a complicated
system of forces to explain his facts of observation, and
when Newton tried a flat space he had to do the same.
Just as the traveller's friend shewed him a better kind of
map, so Einstein has shewn us a better kind of map.
Using this new kind of map, Einstein has been able to
dispense with gravitational pulls, and at the same time
draw a far simpler and far more accurate picture of nature.

Shortest Courses

Although the concept of a curved map of nature first
entered science with Einstein, the complementary concept,
of material objects and rays of light following the shortest
possible course, goes back to the earliest ages of science,

and has a most respectable pedigree behind it. Revolutionary though Einstein's new theory seemed when it was first announced, it merely put science back on the road it had been travelling for two thousand years before Newton. This may not be of much interest to the practical scientist except as a matter of history, but it is of considerable interest to one who wishes to understand the philosophical implications of modern science.

Three hundred years before Christ, Euclid had defined a straight line as the shortest distance between two points, and announced that light travelled in straight lines. Thus he knew that light took the shortest path from point to point—at any rate under ordinary terrestrial conditions. He also knew that light could be deflected from its path by a mirror, and discovered the laws it obeyed when this happened—it followed the same path as a perfectly hard ball bouncing off the mirror.

Euclid saw these only as two totally disconnected facts. About a century later, Hero of Alexandria combined them in a very significant synthesis, shewing that even when light was reflected at a mirror, its path was still the shortest by which it was possible to travel to the mirror and back again. Mirror or no mirror, light followed the shortest path.

Light can experience other accidents besides reflection. For instance, we can see the setting sun after it has passed well below the horizon in the geometrical sense. We say we see it by rays which are "refracted", or bent, by the earth's atmosphere. These rays cannot be following the shortest path, for a bent path can never be as short as a straight one.

Actually, they are following the quickest path. When light passes through any material substance, such as glass or air, the particles of matter slow down its motion, and the denser the substance, the greater the slowing down. Thus when light has to travel through a number of different substances, it can often save time by travelling through a less dense substance even though it has to travel an additional distance to get to and from this substance—just as the steamer saves time by going farther north than either its starting-point or its port of arrival, and taking advantage of the short degrees of longitude up north. As an example of this, the quickest path for a ray from the setting sun to us is one which avoids travelling overmuch through the dense air near the earth's surface, the rays bending round this rather than travelling directly through it. It is found to be a quite general law of nature—known as Fermat's principle—that when rays of light have to make their way through a retarding substance, they take the quickest path from point to point.

Hero had stated his law in the form that light, whether reflected or not, takes the shortest course from point to point. He might have said with equal truth that it takes the quickest, and had he done so, his statement would have been true for refracted light also. Indeed it would have been true for light moving in all the ways known to science; all known light takes the quickest path from point to point.

Interference

The undulatory theory, which interpreted light as waves in an undulating ether, provided a very simple explanation of this. When waves or ripples travel over a pond,

points on the surface of the water are alternately depressed
and elevated beyond the normal level of the undisturbed
water. When two sets of waves are traversing the surface
at the same time, one may tend to elevate a particular
spot on the surface, while the other tends at the same
moment to depress it. When the two effects neutralise one
another in this way, the two sets of waves may jointly
produce less disturbance than either would produce alone.
This is the phenomenon known as "interference"; its
essence is that the disturbances produced by sets of waves
must be treated as algebraic quantities—elevations (positive) and depressions (negative)—and not as mere arith-
metical quantities, and of course the sum of two large
algebraic quantities may be small, or even zero, if they
are of opposite signs.

It is easily shewn that two waves which started from the
same source at the same instant, and so have travelled
for the same length of time, will be always in step with
one another—or, to use the technical expression, in the
same "phase". This will also be the case if their times of
travel differ by the times of one, two, three or any exact
number of complete oscillations. Now if two waves which
are in the same phase, or nearly in the same phase, meet
at any point, their joint effect is greater than that of
either singly—we may, for instance, have two crests superposed, or two troughs superposed; in either case the
disturbance is intensified. On the other hand, if the waves
are in opposite phases, crest is superposed on trough, or
trough on crest, and we have destructive interference.

When waves are sent out from any specified point, we
can imagine them travelling in circular ripples until some-

thing occurs to disturb their regular forward motion; they may, for instance, encounter an obstacle. As soon as this occurs, it is convenient to think of each little disturbance as itself forming a centre for new waves; these new or "secondary" waves now spread out from every point of the primary waves, the whole complex system crossing and recrossing one another and either reinforcing or destroying each other by interference at every point.

Even when no such obstacles exist, we are still free to imagine each wave breaking up into a multitude of new waves at each point of its journey, so that the whole of the disturbed region can be thought of as filled with waves crossing and recrossing one another. Mathematical analysis shews that these waves will reinforce one another all along the path of quickest journey, while they neutralise one another along all other paths. It is not strictly true that the waves travel only along the quickest path; they travel along all possible paths, but destroy one another on all paths except the quickest.

This is not quite the whole story, since we have seen that waves whose times of travel differed by exactly one, two, three, or any integral number of complete oscillations, would also reinforce one another when they met, and so ought to be visible, although neither had travelled by the quickest path. The undulatory theory had the great triumph of its life when this possibility was found to account exactly and completely for all known phenomena of diffraction and interference. These phenomena, which had dealt a fatal blow to the corpuscular theory, seemed to provide incontrovertible proof of the truth of the un-dulatory theory.

Least Action

It was also found that material objects could be brought under the same synthesis as rays of light. Aristotelian doctrine had asserted that substances tended to rise or sink, according as they were light or heavy; every object moved so as to find its own proper place in the ordained scheme of nature. Galileo and Newton made it clear that this was not a universal law, but a mere local effect resulting from the gravitational pull of the earth. In the absence of such forces all objects moved in straight lines with uniform speed; like light they took the shortest path from point to point, or, if their speed of motion was assigned, they took the quickest path from point to point—again like light.

When forces were in operation, their effect was, in Newton's words, to "draw bodies off from their rectilinear path", in which event it was clear that their path could neither be the shortest nor the quickest possible.

The French mathematician-philosopher Maupertuis argued that even in such cases the path must exhibit some perfection worthy of the mind of God. When there were no forces in action, it was already known that this perfection took the form of either the distance or time of the motion being a minimum; hence when forces were in action, something else must still be a minimum. Our modern minds find it a strange line of attack, but it succeeded. Maupertuis discovered a quantity known as the "action", which proved always to have the minimum possible value. This quantity is associated with the motion either of a single object or of a group of objects; just as each bit of travel on a railway involves a certain expenditure in railway fare, so each bit of motion involves a certain expenditure of "action". Our expenditure on railway fares is not usually exactly proportional to either the time or the distance we travel, and in the same way, when forces were in operation, the expenditure of action was not exactly proportional to either the time or the distance of the journey. Yet it was easily calculable, and Maupertuis shewed that objects invariably moved in such a way as to make the total expenditure of action a minimum.

When no forces were in operation, the expenditure of action was exactly proportional to the time, and the new principle of least action absorbed the older principle of least time. Thus the new principle provided a synthesis to cover the motion both of objects and of light—of matter and radiation, the two constituents of the physical universe.

Incidentally, this principle enables us to take a first step at least towards understanding the puzzle of the apparently dual natures of radiation and matter, both of which sometimes remind us of waves and sometimes of particles.

We have already seen that any picture we may draw of nature must be built up of concepts already existing in our minds. The number of such concepts is very limited, but waves and particles happen to be two of the most familiar of them, with the result that we tend to think in terms of waves and particles. For a time it seemed as though radiation could be pictured quite perfectly as consisting of waves, and matter as consisting of particles. We now know that nature is not as simple as this; neither matter nor radiation can be pictured either as pure particles or as pure waves. A wider view shews us both radiation and matter as entities whose behaviour conforms to the mathematical principle of least action. Thus to obtain a true picture of nature, we must try to picture both matter and radiation in terms of familiar things whose behaviour also conforms to this principle, but we again find our choice almost limited to particles and waves. Thus we picture radiation sometimes as particles and sometimes as waves, and also, as we shall see later, do the same with electrons and protons.

Least Interval

Nevertheless a reservation must be made—the simple synthesis of "least action" did not provide a perfect explanation of nature. At first it was found possible to alter and extend it, so as to bring one new phenomenon after another under its scope, but—ominous sign!—each extension made it more intricate and, to all appearances, more artificial, until finally it broke loose from the facts altogether; nothing could make it fit. Even in its most intricate form, it still predicted that gravitating bodies should bend starlight only half as much as they are observed to do, and that the perihelion of Mercury should stand still instead of advancing round the sun at about 80 miles a year.

Then the theory of relativity came to the rescue, first explaining why the principle failed, and then shewing how it could be put right. The restricted theory, described in our previous chapter, shewed that the principle was bound to fail. For any true picture of nature, or principle to explain the workings of nature, must permit of representation in the undivided four-dimensional continuum. The principle of least action, on the other hand, did not permit of representation in this framework until it had been divided up into space and time.

There was found to be only one entity capable of representation in the new framework which could possibly replace action. This was the "interval", the blend of space and time which lay between two events, so that the only minimal principle with any self-consistent meaning is one of "least interval". It was the essence of the generalised theory of relativity that this interval must be measured in a curved continuum. The world-line of a particle or other moving body could be obtained—like the steamer tracks in our earlier analogy—by stretching a tight string from point to point.

When the principle is amended in this way, it takes upon itself the rôle which at one time seemed to be filled by the principle of least action, and is found to govern and to predict the whole motion of the universe, in so far as this is determined by what we used to describe as the forces of gravitation. It seems possible, although by no means certain, that electrical forces admit of explanation in terms of the same principle, so that when an electron appears to be compelled to describe a curved orbit by electric forces, it also is finding the shortest possible path through a curved continuum. Einstein's recent "Unitary Field-theory" attempts to specify the exact kind of continuum necessary for such an explanation, but its success is not yet established. If ever complete success is achieved in this direction, the principle will equally govern the motion of a ray of light and of a moving body, and will remain valid whatever physical agencies are in action, so that we shall be able to combine all the operations of nature in one synthesis; they will all have become shortest courses in a curved four-dimensional space.

Generalised Relativity

In this way, Hero's first simple synthesis of the two laws of Euclid has been gradually extended and modified until it has finally emerged as a general principle covering all the large-scale phenomena of nature, and possibly sub-atomic phenomena as well. We used to think of the principle as applying to particles of matter and to rays of light existing and travelling in a framework of space and time. But in the process of making a perfect fit between the principle and the observed facts of nature, we have had to discard space and time as objective realities, forces and mechanism have dropped out of the picture altogether, and we are left only with empty space and empty time, first welded together to form a four-dimensional continuum differing in quality from either space or time, and then curved and contorted. We can no longer think of the varied phenomena of nature as arising from a blind dance of atoms as they are pushed and pulled about by mechanical force; we must attribute them to efforts of we know not what to find the shortest path through the tangled maze of the space-time continuum.

At this point it becomes natural to inquire whether Einstein's picture represents anything in ultimate reality, or merely provides a convenient way of describing phenomena. For we must remember that the most convenient description is not always that which is closest to reality. Although ships' captains are aware that the world is round, they still find it most convenient to map out the

tracks of their ships on flat Mercator projections, as though
the earth was flat. In the same way, Einstein's straight
paths in a curved space may conceivably be merely convenient pictures which represent the phenomena but not
the reality behind.

The Einstein Universe

Mathematical analysis shews that there are more ways
than one of curving the continuum so as to explain the
paths of astronomical bodies and rays of light. These all
give equally good pictures of the astronomical phenomena
of the solar system, or any other small part of the universe,
yet one at most can represent ultimate reality. Many are
disqualified because they lead to obvious absurdities when
the universe is considered as a whole. Einstein found one
way which he considered free from such objections.

This postulated two distinct kinds of curvature. The
first was a curvature inherent in the continuum itself,
which rolled the whole continuum up into a closed
surface, much as the whole surface of the earth is rolled
up into a closed globe. The second consisted of local
irregularities which were superposed on to the main curvature, just as a curvature of hills, mounds and molehills
may be superposed on to the main curvature of the earth's
surface. These smaller irregularities were caused by, or at
any rate associated with, the presence of matter, and were
responsible for the observed curvatures of the paths of
planets and rays of light. Over a small region of space,
such as the solar system, the main curvature produced too
small an effect to permit of observation.

If the continuum were curved in this way, then space,

being a cross-section of the continuum, was also curved. It was moreover curved in such a way that there was only a finite amount of it. This representation possessed certain definite advantages over all older views of space, which had always been confronted with the dilemma that, although it was impossible to imagine any limit to space, yet unlimited space was objectionable on purely scientific grounds. If matter extended through unlimited space, there would be an infinite amount of it exerting its attraction on planets, stars and galaxies, and this would cause them to move at speeds far greater than those actually observed—at infinite speeds, in fact. The only escape would be by supposing that there was only a finite amount of matter, and as this could only occupy a finite amount of space, it left an infinite amount of space entirely devoid of matter. Such a concept could not be disproved as being in any way ridiculous or impossible, but it was certainly not convincing by its inherent reasonableness. Kant had dismissed it on the grounds that an infinite empty space would contain nothing by which to locate the position of a finite material world. If the question "Where is the finite matter in infinite space?" admitted of no answer, then there could not, according to Kant, be finite matter in infinite space.

However serious these difficulties were, Einstein removed them all by his concept of a closed finite space.

He found that, when the average density of matter in space is assigned, there is one and only one radius at which space can stand in equilibrium without either expanding or contracting. He accordingly supposed this to be the actual radius of space; it could of course be calculated as soon as the average density of matter in space was discovered observationally.

Since space was supposed to retain the same radius through all time, the curvature of the space-time continuum could not be geometrically like that of the surface of the earth; it was rather like that of a roll of paper, or better still that of a single sheet of paper pasted so as to form a cylinder of paper—like a postal-tube (cf. fig. 3). In this model of the continuum, any cross-section—the paper itself, not the circular area it encloses—represents space at any instant, while the passage of time is represented by length-wise motion along the cylinder. Thus space became finite and constant in amount, while time remained infinite, extending from an eternity back in the past, through the present to an eternity in the future.

The Expanding Universe

Recent mathematical investigations have shewn that the continuum cannot be represented by so simple a model. Einstein originally introduced his large-scale curvature into the universe to keep it in equilibrium. Friedmann, Lemaître and others have shewn that a universe whose equilibrium was secured in this way would not stay in equilibrium. It would be unstable, in the sense that space would at once start expanding or contracting—the general mathematical theory leaves both alternatives open—and that the process would continue with ever-increasing speed. Thus we must not picture space-time by a cylindrical roll of paper, but rather by a cone or horn-shaped surface, such as the cardboard surface of a megaphone (cf. fig. 4). Time is still represented by the central axis. Space, the cross-section of the horn, is still finite, but for ever changes its dimensions as we move about in time. In other words, space cannot remain of constant size, as Einstein originally imagined, but must be for ever expanding or contracting.

If Einstein's molehill curvature represents anything in ultimate reality, and is not a mere convenient means of

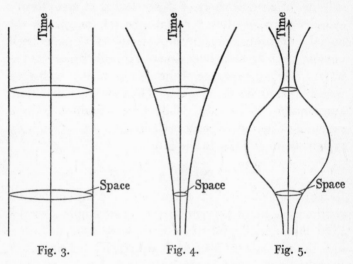

Fig. 3.   Fig. 4.   Fig. 5.

Diagrammatic representations of space-time to exhibit various theoretical possibilities.

picturing the paths of the planets, then the large-scale curvature, which follows almost as a logical corollary to it, ought also to represent something in ultimate reality. Clearly it is important to look for observational evidence of its existence.

The most obvious property of this large-scale curvature is that it closes space up, so that if we tried to travel on for ever through space, we should merely come back to our starting-point, as Drake did when he circumnavigated the globe. It is of course no good our trying to obtain a proof of the curvature of space by an actual circumnavigation of space—for one thing life is too short. A ray of light might have a better chance, for it travels at ten million miles a minute and is not limited to a lifetime of three-score years and ten. It was at one time thought that a sufficiently powerful telescope might enable us to look round space and see our own galaxy by light which, starting many millions of years ago, had travelled round the whole of space, and finally come back to its starting-point. Such an experience would of course constitute a very direct and convincing proof of the curvature of space, but we no longer believe it to be possible.

It is not, however, necessary to travel round space to obtain a convincing proof of its curvature, any more than it is necessary to travel round the earth to prove that its surface is curved. If we draw a circle on a perfectly flat piece of paper, we know that its circumference is π times its diameter, where π denotes the number 3·14159.... This is true, no matter how large or how small our circle may be, provided always that we draw it on a perfectly flat surface. It is not, however, true of a circle drawn on a curved surface such as that of the earth. A circle of small diameter still has a circumference equal to 3·14159 times its diameter, but the ratio becomes less as the circle is made larger. A circle of 1000 miles diameter does not have a circumference of 3141·59 miles, but only of about 3110 miles. If a surveyor were to draw such a circle on the earth's surface, and then measure its circumference, he would obtain a ready proof of the curvature of the earth's surface.

In theory it is possible to test the curvature of space in a similar way. If we construct a small sphere of any substance, its surface will be π times the square of its diameter, where π is still the same number 3·14159. Now if space were uncurved, the surface of any sphere would always be 3·14159 times the square of its diameter, no matter how great this diameter might be. But if space is curved, the ratio continually decreases as the sphere gets larger, just as on the earth's surface the ratio of circumference to diameter decreases as the circle gets larger.

If then we could map out an immense sphere in space and measure up the total area of its surface, we should have an immediate means of testing whether space is really curved in the way that Einstein imagined. Yet even if the curvature exists, its scale is so large that its effects are inappreciable in the solar system, and we should have to make a sphere millions of millions in diameter before we could hope to detect it. And, apart from the practical difficulties of mapping out a circle of such dimensions, there are two theoretical difficulties—first (p. 76) we have no means of locating points in space, and second (p. 72) we have no objective means either of drawing straight lines or of measuring their lengths.

Although this line of thought will not enable us to test whether space is curved, it goes some way towards helping us to imagine the kind of curvature postulated by the generalised theory of relativity—space contracts as we get farther from home, so that the content of a sphere of assigned radius is always less than it would be if space were flat. We

can construct a flat area in our imaginations by joining a
number of triangles together at their vertices, as in fig. 6,
but if we want to imagine a curved (say a spherical)

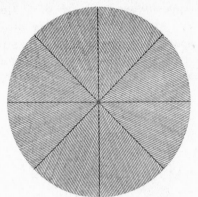

Fig. 6.

Fig. 7.

surface, we must replace our triangles by areas shaped like
the leather sectors which are stitched together to make a
football, as in fig. 7. In much the same way we can

imagine a flat space formed by joining a number of sugar-cones together at their vertices, but if we want to form a curved (spherical) space, we must replace our sugar-cones by spindle-shaped bodies, as in fig. 8. If the reader can imagine enough spindles tied together at *A* to fill the whole of the space surrounding *A*, and then (this is where the difficulty comes) imagine them all bent about, *equally and similarly*, until all their other ends *B*, *B'*, *B''* meet in

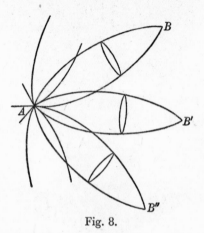

Fig. 8.

a point, he will have made for himself a sort of mental picture of spherical space. More likely, however, he will be unable to imagine this at all, because of the difficulty of conducting his imagination out of ordinary three-dimensional space; he will then have a proof of the impossibility, to which we have so often referred, of either picturing or describing things except in terms of concepts made familiar by our everyday life.

As the curvature of space cannot be directly tested in

either of the geometrical ways we have just described, we must fall back on the more indirect way of examining whether the various mathematical consequences of this curvature are to be found by observation. If the curved continuum merely provides a convenient means of picturing phenomena, there is no reason why all its mathematical consequences should be found in nature; mere representation must be expected to part company with reality somewhere. On the other hand, if this curved continuum has a real existence in nature, all the mathematical consequences of this existence ought to be confirmed by observation. And the principal of these is that space must be either expanding or contracting at a uniform rate throughout its whole extent.

Now the great nebulae out beyond the Milky Way provide just the means of testing this prediction of theory. They are the largest and most distant objects known to astronomy, and yet, in relation to the universe as a whole, they are mere straws floating in the stream of space, and ought to shew us in what way, if any, its currents are flowing. If the continuum is curved in the way we have described, these nebulae ought all to be receding from us, or else all rushing towards us, the speed of each nebula being exactly proportional to its distance from us.

At this point observation takes up the tale. If recent astronomical observations can be taken at their face value, these nebulae are all receding from us, and this at quite terrific speeds. Moreover, their speeds are almost precisely proportional to their distances, exactly as demanded by theory. Nebulae whose light takes a million years to reach us are receding at (in round numbers) 100 miles a second, nebulae at twice this distance at double this speed, and so on. Nebulae whose distance is estimated to be 135 times as great as this—so that their light takes 135 million years to reach us—have just been found to be receding from us at the colossal speed of 15,000 miles a second, the greatest speed so far known to astronomy.

These speeds are so great that many astronomers have doubted whether they are real—surely, they say, the observations must permit of some other and less sensational interpretation. It may be so; we are still a long way from being able to pronounce a final judgment on these questions.

Sir Arthur Eddington* has recently tried to investigate, in a purely theoretical manner, the speeds with which the nebulae ought to move if the universe were expanding in the way required by the theory of relativity. The speeds he calculates agree with those actually observed to within a factor of about 2, which is as good an agreement as could reasonably be expected. The whole investigation is extremely speculative and does not yet, I think, command the general assent of mathematicians. If ever these calculations can be put beyond criticism, they will provide a very strong confirmation of the whole theory of the expanding universe, as developed by Friedmann and Lemaître.

On the other hand, there are very grave astronomical objections to accepting the observed speeds of recession as real. If they are real, the universe must be changing very rapidly; it is doubling its dimensions every 1,300 million years or so. If we assume that the speeds have always been

* *The Expanding Universe* (1933), chap. IV.

as at present, and trace the motion back for 2,000 million years, we find the whole universe concentrated in a quite small region of space. Actually the theory of the expanding universe shews that the speeds would diminish as we go backwards in time, and that there is no definite limit to the time during which expansion can have been in progress, but it also suggests rather forcibly that this time can hardly be more than about a hundred thousand million years. Against this, the time needed for the universe to attain its present stage of evolution can be estimated in a great number of ways, and all agree in indicating a period of millions of millions of years. It is exceedingly difficult—although perhaps not absolutely impossible—to imagine that the universe can have been evolving for ten or a hundred times longer than space has been expanding. It is even more difficult—although again perhaps not absolutely impossible—to imagine that space can have been expanding for millions of millions of years. The difficulty is so grave as to cast real discredit on the whole mathematical theory of the expanding universe.

A recent short note by Einstein and de Sitter may be found to contain a means of escape from this very serious dilemma. We have seen how Einstein originally thrust his large-scale curvature on to the universe because he saw no other way of keeping it quiet, and restraining it from either exploding or collapsing. Now that space appears to be exploding in spite of all Einstein's efforts to save it, the inherent large-scale curvature seems to play a less essential part in the scheme of things than it once did.

Einstein and de Sitter have accordingly examined what reasons, if any, remain for supposing that space possesses this inherent curvature. They find none at all. It is no longer needed to keep space at rest, because space is not at rest, and neither the mathematical equations nor the observed recessions of the nebulae in any way require it. Thus we become free to suppose that space would be flat if it were perfectly empty of matter, and that it owes the whole of its curvature, both coarse and fine, to the objects which occupy it. While this does not carry us much further towards a positive understanding of the nebular motions, it brings a whole new class of possibilities into the field. It has, for instance, been suggested that the universe may be undergoing a succession of alternate expansions and contractions (cf. fig. 5, p. 132); this would account for the observed recessions of the nebulae, and yet give us all the time we want for the evolution of the universe; there is no longer any conflict with the general evidence of observational astronomy.

We must not regard any of the foregoing speculations or conclusions as in any way final or established. Indeed, science is only just entering upon its latest and most comprehensive problem—the study of the universe as a single entity—and it would be folly to treat the first tentative results as final. Yet, although these can hardly be said to have led to definite conclusions so far, they nevertheless hold out hope that conclusions may not be very distant. And they illustrate once again that it is usually the totally unexpected that happens in science—the unaided human mind can seldom penetrate far into the darkness which lies beyond the small circle of light formed by direct observational knowledge.

Even the meagre results so far obtained seem to shew

that nature is one and not many. The different sciences have each drawn their own pictures of small fragments of nature which form their special objects of study, and we now find that these fragmentary pictures piece together to form a consistent whole. An experiment performed by two physicists, Michelson and Morley, with a view to measuring a time interval of less than a million millionth part of a second, or a length of less than a thousandth part of an inch, led, through the theory of relativity, to a picture of the whole vast universe, which depicts it as exploding like a burst shell, its most distant objects unanimously rushing away from us. We examine these objects through our largest telescopes and find that they are, to all appearances, rushing away in precisely the way predicted by theory.

The Nature of Space

The point which is of immediate interest to our present discussion is the following. Unless this apparent agreement between theory and observation is wholly illusory, it provides us with evidence of a contact between the theory of relativity and reality at the furthest point to which this theory has so far been pushed. It suggests very strongly—although of course it does not prove—that the curved continuum postulated by this theory has more reality than that of a mere convenient explanation of the apparently curved paths of planets and cricket balls, just as the curved surface of the earth has more reality than that of a mere convenient explanation of the apparently curved tracks of steamers.

Even so, it only tells us of the metrical properties of space, and nothing as to its essential nature. Indeed, there would appear to be little advantage in discussing this latter problem. After 2,000 years of metaphysical discussion, the question stands much as Plato left it in the *Timaeus* (pp. 74, 146); the growth of scientific knowledge has done little more than negative the speculations of subsequent philosophers. Of all external entities, perhaps space is the one whose essential nature is least likely to be understood by the human mind, since it is hardly probable that what is completely external to the mind, and without effect on the mind, will admit of being pictured in terms of familiar concepts inside the mind.

Although the new curved continuum is still a blend of space and time, these constituents no longer enter it in similar or even symmetrical ways. In our simple diagrammatic analogy, space was the cross-section and time the axis of a cone, and however much new knowledge may change the details, some such distinction seems likely to persist. Such a continuum does not satisfy the invariant condition, which was found to be essential in the restricted theory, of giving a picture which does not suffer by being hung askew. It therefore contains in itself a unique mode of separation into space and time, which we may now designate as absolute space and absolute time. The restricted theory of relativity, which we discussed in our previous chapter, shewed that any division into space and time was subjective in respect of such phenomena as we could observe and measure in our laboratories. The generalised theory which we are now discussing suggests that just as our individual consciousnesses recognise a sharp and clear-cut distinction between space and time, so also does nature on the grand scale. This distinction, which we first find in our own minds, vanishes for a time when we study objective nature on the small scale, but apparently reappears in the cosmos as a whole.

Neither the mathematical theory we have just described, nor the interpretation of the astronomical observations, are sufficiently certain to warrant the drawing of any conclusions except as almost random conjectures, but a simple analogy may suggest the kind of conjecture that presents itself.

Let us, very unpoetically, compare the human race to a race of worms living inside the earth, and capable of burrowing about in it but never reaching its surface. As their bodies would be subject to gravitational forces, their minds would be conscious, through their nervous systems, of a distinction between horizontal and vertical directions inside the earth. They would not be able to pick out an absolutely permanent and unaltering horizontal and vertical, since a worm who was moving with an acceleration would experience a different horizontal and vertical from his fellows who were not. In spite of this, the worms might still feel sure that the horizontal was somehow essentially different from the vertical. Suppose now that they took to science and built laboratories, still inside the earth, in which to study electromagnetism and optics. They would be unable to detect any distinction between horizontal and vertical in their laboratories, because the laws of electromagnetism and optics treat all three directions in space equally. If, then, they knew of no sciences but these, they would be unable to discover any scientific justification for their intuitive feeling that horizontal and vertical were really dissimilar. Finally, to come to the climax, one of them might burrow his way out to the surface of the earth and discover that their intuitive feeling was based on a real fact of nature. He would then see that nature contained something more than the sciences they had studied in their laboratories, and would realise that they had all the time been in contact with nature through this something more.

In the same way, our minds are conscious of a radical distinction between space and time which does not appear to extend to physical phenomena; these seem so similar in the continuum and so dissimilar when apprehended by our minds. Through our consciousnesses, we break up the space-time product into space and time, while electrons and protons and radiation cannot. If this distinction is ultimately found to be real, as our present vague and uncertain knowledge seems to suggest, we may be tempted to conjecture that our minds are in contact with reality through other than purely physical channels.

Finally, we may notice that, if a more complete knowledge of the continuum as a whole is ultimately found to restore a meaning to absolute space and absolute time, the problems which were indicated at the end of Chapter III do not arise.

In that chapter we pointed out how the concept of the space-time continuum—neither space nor time being complete in themselves, and only acquiring objective reality when blended into a single whole—was in accordance with a view which certain metaphysicians had taken of space. In the present chapter we have considered the properties of this blend of space and time in more detail. We have seen that, according to the picture drawn by the generalised theory of relativity, space must be finite in amount, and must possess a texture, defined by the different curvatures at its various points, so that it is in some way differentiated from mere emptiness. Again, these qualities satisfy the requirements of the metaphysician. Writing twenty years before the generalised theory of relativity appeared, Bradley described these in the following words: *

"Empty space—space without some quality (visual or muscular) which in itself is more than spatial—is an unreal abstraction. It cannot be said to exist, for the reason that it cannot by itself have any meaning. When a man realizes what he has got in it, he finds that always he has a quality which is more than extension. But, if so, how this quality is to stand to the extension is an insoluble problem".

And again, with reference to finite space:

"For take space as large and as complete as you possibly can. Still, if it has not definite boundaries, it is not space; and to make it end in a cloud, or in nothing, is mere blindness and *our* mere failure to perceive. A space limited, and yet without space that is outside, is a self-contradiction. But the outside, unfortunately, is compelled likewise to pass beyond itself; and the end cannot be reached. And it is not merely that we fail to perceive, or fail to understand, how this can be otherwise. We perceive and we understand that it cannot be otherwise, at least if space is to be space. We either do not know what space means; and, if so, certainly we cannot say that it is more than appearance. Or else, knowing what we mean by it, we see inherent in that meaning the puzzle we are describing. Space, to be space, must have space outside itself. It for ever disappears into a whole, which proves never to be more than one side of a relation to something beyond".

* *Appearance and Reality*, pp. 37, 38.

This quotation raises a metaphysical dilemma which science alone cannot claim to solve. If the whole continuum is finite, what can there be outside the continuum except more continuum?—which proves that our original continuum was not the whole continuum. And how can space be expanding, since there is nothing for it to expand into except more space?—which proves that what is expanding cannot be the whole of space, and so on. We shall return to this in a later chapter.

Finally, those who hold that "out of Plato come all things that are still debated among men of thought" may be tempted to claim that Plato anticipated Einstein in evolving the whole of nature out of the metrical texture of space. They may even claim that he anticipated Friedmann and Lemaître in respect of the instability of the Einstein universe. For he wrote*:

"Even before the birth of a heaven, there were these several three—being, space, becoming. Hence as the foster-mother of becoming† was liquefied and ignited and received the shapes of earth and air and underwent further affections consequent on this, she took on many motley guises. And since the forces with which she was filled were neither alike nor equipoised, there was no equipoise in any region of her; she was swayed and agitated with utter irregularity by these her contents, and agitated them in turn by her motion."

* *Timaeus*, Taylor's translation, p. 52.
† *I.e.* Space.

CHAPTER V

THE TEXTURE OF THE EXTERNAL WORLD—
MATTER AND RADIATION

We must now leave the vastness of astronomical space, to pass to the other extreme of the scale of size and explore the innermost recesses of the ultra-microscopic atom. While the phenomena of astronomy may shew us the nature of space and time, it is here, if anywhere, that we may hope to discover the true nature of matter and of material objects, the contents of space and time.

The Structure of Matter

We have seen how the atomic concept of matter gradually gained scientific recognition, and finally appeared to be securely established when Maxwell and others shewed that a gas could be pictured as consisting of hard bullet-like atoms or molecules flying about indiscriminately at speeds comparable with those of ordinary rifle bullets. The impact of these bullets produced the pressure of the gas; the energy of their motion was the heat-energy of the gas, so that heating up the gas resulted in its bullets travelling faster; the viscosity of a gas was caused by the drag of one bullet on another on the rare occasions on which actual collisions occurred, and so on. These concepts made it possible to explain a great number of the observed properties of gases, both qualitatively and quantitatively, with great exactness. Yet a residue obstinately defied explanation, and it is only recently that an explanation of these has been obtained, in terms of new and very different concepts to which we shall shortly pass.

This picture of matter as consisting of hard indivisible atoms had to be modified when Sir J. J. Thomson and his followers began to break up the atom. They shewed that the atom was far from indivisible; small fragments of it could be knocked out by bombardment, or pulled out by sufficiently intense electric forces. These fragments proved to be all similar—the electrons. They all had the same mass, and carried the same electric charge, which was conventionally described as being of negative sign. It was subsequently found that the remaining ingredients of the atom were also similar electrically charged particles—the protons. Their charges were opposite in kind to the charges on the electrons, and so were described as positive in sign.

There were many reasons for supposing all the atomic constituents to be of minute size. For instance, it was found that radio-active substances shot off two kinds of projectiles, a less massive kind known as β-particles, which proved to be rapidly moving electrons, and a more massive kind known as α-particles, which proved to be identical with the central nucleus of the helium atom. This is known to consist of four protons and two electrons. When particles of either kind were shot at matter they penetrated it to a considerable depth, which suggested that they were of very small dimensions. A tennis ball weighs about the same as a rifle bullet, yet if we fire both at the same piece of wood, the bullet will penetrate a considerable distance, because its mass is concentrated in a very small space, while the tennis ball will not penetrate at all; both α- and β-particles were found to behave like rifle bullets rather than like tennis balls. Not only so, but when they were fired at a thin film of metal, the majority passed through without being substantially deflected from their courses, which seemed to shew that the electron and protons of the metal film were themselves of minute size. Thus there appeared to be fairly conclusive evidence that the ultimate ingredients of matter were of the nature of small particles carrying highly concentrated charges of electricity.

It has never been found possible to measure the sizes of these particles directly. It is often supposed that the diameter of the electron must be about 4×10^{-13} cms.; it cannot be less than this, for if the electrical charge of the electron were compressed into any smaller volume, the inertia resulting from this alone would necessarily be greater than that of the total observed mass of the electron. Yet this raises a serious difficulty. According to the generally accepted theory, the nuclei which form the centres of the most massive atoms, such as gold or uranium, must contain a large number of electrons as well as protons. These nuclei are, however, themselves so small in size that they could not contain the requisite number of electrons of the size just mentioned inside them, even if the protons occupied no space at all. This shews that the concept of electrons and protons as small charged particles is at best only a picture, and a picture which cannot be true to nature in all particulars. Nevertheless, a small spherical particle of radius 2×10^{-13} centimetres and charged with $4 \cdot 777 \times 10^{-10}$ units of electricity reproduces many of the properties of the electron, and probably no one has ever regarded it as providing a complete picture which was true in all particulars.

Radiation

As regards their material structure, all objects are built up of these two kinds of electrified particles, but they contain also the intangible constituent of energy, which may be set free from all association with matter, when it travels through space in the form of radiation. We have seen how, throughout the nineteenth century, radiation was pictured as waves in the ether. This picture not only failed to describe the propagation of radiation in ways which have already been described; it also failed to account for some of the most fundamental properties of the radiation itself.

We know how a pendulum swinging in air continually loses energy to the molecules of air which impinge on it; unless it is kept in motion by clockwork it soon comes to rest, the energy of its motion being transformed into waves of the surrounding air which are subsequently dissipated into heat. In the same way a steamer soon comes to rest when its engines are stopped, the energy of its motion being used in setting up waves in the surrounding sea. And, again in the same way, it can be shewn that if material bodies were surrounded by a sea of ether, their energy would be rapidly dissipated in setting up waves in the ether. Calculation shews that this process would continue until the material bodies, like the pendulum and the steamer, had no energy left at all; their whole energy would have passed into the ether, where it would take the form of radiation of very short wave-length. This is true of all kinds of energy, so that a hot body ought speedily to lose all its heat-energy to the ether, and fall to the absolute zero of temperature.

Instead of this, experiment shews that a state of equilibrium is soon attained in which a body receives back from the surrounding space exactly as much radiation as it pours out into it. For instance, disregarding certain small internal stores of heat, the average temperature of the earth is such that it loses just as much energy by radiation into space as it receives back from space in the form of solar light and heat. If the earth were suddenly made hotter than it now is, it would cool down to its present temperature, but not to the absolute zero; if it were made cooler, it would warm up to its present temperature. To take a more precise case, if a heating system maintains all the walls of a closed room at exactly 60° F., then every object in the room will stay permanently at exactly 60° F.—this is why we can say that a thermometer gives "the temperature of the room".

In such a state of equilibrium, every object gives out just as much radiation as it receives. In the idealised case of an object which has no reflecting power at all, and so appears perfectly black (p. 19), the radiation is known technically as "black-body radiation", and is said to have the temperature of the object which emits it. Radiation of this kind can be analysed into its different constituents with great accuracy, and its quality is found to be as unlike as possible to what it would be if radiation consisted of waves in a substantial ether.

Quanta

In the last years of the nineteenth century, Planck tried to discover the reason for this divergence, and, just as the century was closing, he put forward the ideas out of which the vast structure of the quantum theory has since arisen. He shewed that it was possible to account exactly for the observed state of equilibrium between matter and radiation, by the assumption that radiation was atomic in its nature. He supposed it to occur only in complete multiples of a unit which he called the "quantum". This unit was not the same in amount for all kinds of radiation, but depended on the wave-length of the radiation, and so also on its period of oscillation or its "frequency"—the number of oscillations performed in a second. To be precise, radiation which oscillated ν times a second was supposed to occur only in complete units of energy of amount $h\nu$, where h was a quantity, now universally known as Planck's constant, which is found to pervade the whole of atomic physics. Thus blue or violet light, being of high frequency, consisted of quanta of great energy, while red light, which is of low frequency, consisted of quanta of small energy. The greater the energy of the quanta, the greater their capacity for producing atomic change. This is why blue light causes pigments to fade and affects photographic plates, where red light is ineffective.

The recently-discovered X-radiation was known to be of enormously high frequency, so that on Planck's theory its quanta ought to possess exceptionally great energy. It was soon remarked that when this radiation was passed through a gas, a few of the molecules of the gas were shattered, but the vast majority remained entirely unaffected by the passage of the rays. Had the rays consisted of waves travelling through an all-pervading ether, it might reasonably have been expected that they would treat all the molecules they encountered in the same way, or at least in approximately the same way; actually, less than one molecule in a billion seemed to be singled out for destruction. It was further found that doubling the intensity of the radiation did not double the damage done to each molecule, but merely doubled the number of molecules that were damaged. This was subsequently found to be true of radiation of all kinds, including ordinary visible light. It was exactly what was to be expected if radiation consisted of small point-like atoms of radiation, like the old Newtonian corpuscles.

Photons

In 1905 Einstein crystallised these concepts and hypotheses in his theory of light-quanta, according to which all radiation consisted of discrete bullet-like units, which he called "light-quanta" at the time, although we now call them "photons". When an atom was struck by a photon, it might be either disturbed or shattered, according to the amount of energy which the photon brought to the attack, and by observing the amount of damage done to the atom, it became possible to calculate the energy of the individual photons. This invariably proved to be exactly one quantum —if the incident radiation which attacked the atoms was of frequency v, the change produced in each affected atom represented an expenditure of energy hv.

One of the fundamental consequences of the theory of relativity is that every kind of energy has mass associated with it. Thus a photon must possess mass of its own, and it is just as accurate to speak of the mass of a photon as of the mass of an atom or of a motor car. As photons are always in motion, we may also speak of the momentum of a photon, much as we speak of the momentum of a motor car, although there is the essential difference that photons always move with the same speed, the speed of light, whereas motor cars move with variable and different speeds.

Professor Compton of Chicago has recently found very direct evidence of the existence of this mass, and has been able to measure its exact amount. When a photon strikes an atom, its energy is not always completely absorbed by the atom; it may occasionally strike a particular electron in an atom and rebound from it like a perfectly hard bullet. In such cases the photon loses part, but not all, of its energy to the electron with which it has collided. Compton found that when this happens to a photon, its frequency changes in such a way that after the collision the energy is precisely h times the frequency of the radiation, as of course it also was before the collision. The circumstances of the recoil made it possible to calculate the momentum, as well as the energy, of the photon, and this proved to be hv/c. This is exactly the amount of momentum which the theory of relativity predicts must be associated with energy hv moving with the speed of light.

These various experiments suggest that radiation may be pictured as consisting of bullet-like units, which travel through space very much like shot fired from a gun and have nothing to do with any supposed ether. In this new picture, the constant speed of 186,000 miles a second at which radiation travels is no longer regarded as the speed of waves; we have instead to imagine that photons of radiation are endowed with inertia, like a bullet or an electron. This inertia keeps them moving in a straight line with a uniform speed, although nothing in this picture explains why this speed should always be 186,000 miles a second. This last fact shews that the particle picture by itself is incomplete.

Quantitatively, the experiments shew that the momentum of a photon is connected with its wave-length by the relation

$$\text{momentum} \times \text{wave-length} = h,$$

while its energy is connected with its period of oscillation by the relation

$$\text{energy} \times \text{period of oscillation} = h.$$

Finally the wave-length and period of oscillation are connected by the relation, which survives from the undulatory theory,

$$\text{wave-length} = \text{period of oscillation} \times c,$$

where c is the uniform speed of light.

The evidence which these experiments provide for the real existence of photons is of the same general nature as that which other experiments provide for the existence of electrons. In each case experiment suggests an indivisible entity having definite quantities associated with it—e and m for the electron, and h and c for the photon—and measurement of these quantities yields uniformly consistent values. No experiment yet performed has suggested that fractions of either entity can exist independently; fractions of a photon are as unknown as fractions of an electron.

The radiation with which we are usually concerned in atomic physics is produced by disturbances or upheavals of single atoms, and it is found to be a general law that every such disturbance produces one, and only one, complete photon. As mass is conserved through a change of this kind, the atom must lose mass exactly equal to the mass of the photon it emits. When the disturbance consists only of a rearrangement of the outermost electrons of an atom, the resulting change of mass is only a few millionths of the mass of a single electron, and the photon has the wave-length of visible light—it is by the entry of such photons into our eyes that we see things. A re-arrangement of the inner electrons of the atom produces X-radiation, in which each photon has a mass of perhaps the 10,000th part of the mass of an electron. If the nucleus of the atom rearranges itself, we have the still more penetrating γ-radiation, in which each photon has a mass comparable with the whole mass of an electron. Finally, the hardest constituent of cosmic radiation, the most penetrating radiation known, has photons of mass about equal to that of a complete atom of helium, while the next most penetrating constituent has photons of mass about equal to that of a hydrogen atom. It is possible, then, that these photons may be produced by the total annihilation of atoms of helium and hydrogen, or, more probably, by the annihilation of electrons and protons to an equivalent extent in more complex atoms.

The Kinetic Theory of Radiation

Just as Maxwell was able to explain many of the properties of a gas by picturing it as a medley of bullet-like molecules, so we can explain many of the properties of radiation by picturing it as a medley of bullet-like photons. The pressure of a gas can be pictured as resulting from the impacts of its molecules, and in the same way the pressure of radiation can be pictured as resulting from the impacts of its photons. And again, just as the energy of a gas is the sum of the energies of its molecules, so the energy of the radiation in any space is the sum of the energies of the photons in that space.

When we picture radiation as consisting of waves, the quantity known as the "intensity" of the radiation is proportional to the energy of the waves at any point, or, even more pictorially, to the "storminess" of a sea of ether. When we picture radiation as consisting of photons, we can no longer interpret the intensity in this way. We may, however, give it a statistical interpretation; we can define it as proportional to the chance of finding a photon at the point in question, just as the density of the gas at a point is a statistical concept, and is proportional to the chance of finding a molecule there. The temperature of radiation, like the temperature of a gas, is also a statistical concept. We cannot speak of the temperature of a single photon, any more than of that of a single molecule. We say that "black-body radiation", which experiences no change either of quality or quantity when it interacts with heated matter, has the same temperature as the matter, but the temperature belongs to the crowd of photons and not to the individuals separately.

Such radiation may be pictured as a crowd of photons moving equally and indiscriminately in all directions, just as a gas in equilibrium may be pictured as a crowd of molecules moving equally and indiscriminately in all directions. The energies of the separate photons conform to the statistical law known as Planck's law, just as the energies of the molecules in a gas conform to Maxwell's law. Various other concepts, such as those of the two specific heats, of their ratio, and of adiabatic changes, mean much the same for the radiation as for the gas, and permit of the same pictorial representation, photons of course replacing molecules.

We may picture the photons as retaining their individual identities through all changes except that of being completely absorbed into, or emitted out of, an atom or a molecule. They may change their energies, but then they adjust their frequencies to their energies so that each photon remains a complete unit. Suppose for instance that "black-body radiation" is darting about inside an enclosure, whose volume can be varied by a cylinder-piston arrangement, and is being continually reflected from its walls. When we compress a gas inside a cylinder, the advancing piston does work against the pressure of the gas, and this work reappears as an increase in the energy of the separate molecules—i.e. as heat. When the cylinder is filled with radiation, the advancing piston does work against the pressure of the radiation, and this reappears as an increased energy of the photons. Let us imagine that we reduce the volume accessible to the radiation to one-eighth, equivalent to an all-round reduction of linear dimensions to half. It can be shewn that each photon will double its energy, and so also will double its frequency and halve its wave-length. Thus wave-lengths and enclosure are uniformly reduced to half-scale, and the new radiation is at double the original temperature. If we suppose the photons to have retained their identity, there are eight times as many per unit volume, so that the density of energy has increased sixteen-fold—as the fourth power of the temperature. This is exactly what is observed, the fourth-power law being known as Stefan's law. And, again as with a gas, the pressure is proportional jointly to the density and temperature, so that this also varies as the fourth power of the temperature, which again agrees with observation.

Just as we can picture "black-body radiation" as a random crowd of photons, so we can picture a beam of radiation as a regular shower of photons, all moving in parallel paths. This, of course, corresponds to a blast of gas in which all the molecules move in parallel paths, their ordinary heat-motion being either neutralised or neglected. On the other hand, a beam of light is in one respect more intricate than a blast of gas, since in addition to its motion through space it possesses the property we describe as polarisation.

The nineteenth-century picture of radiation attributed polarisation to angular momentum of the ether; in our present picture of radiation, we must attribute it to angular momentum in the separate photons which form the radiation. In brief, not only must our radiation move through space like bullets, but each bullet must have a spin like that caused by rifling. Planck's constant h has the same physical dimensions as angular momentum, and we find that we must picture all photons as spinning with the same angular momentum $h/2\pi$, which may be in either direction—right-handed or left-handed. In a beam of circularly polarised radiation, we picture the photons as all spinning in the same direction. If the beam is elliptically polarised, more photons spin in one direction than in the other; if plane-polarised, the numbers are equal. If it is not polarised at all, the proportion of the two kinds continually varies at random, but the laws of probability secure that the actual ratio shall never wander very far from unity. This spin has recently been detected and measured by Raman and Bhagavantam.

Inadequacy of the Particle Picture of Radiation

This picture of radiation as a crowd of bullet-like photons has many advantages, but also suffers from many limitations, which shew that it does not present us with a complete picture of reality, but at best only of certain aspects of reality. We have already mentioned one conspicuous

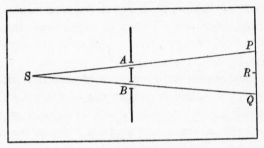

Fig. 9. (This is purely diagrammatic and is not drawn to scale.)

instance of its failure: nothing in the particle picture explains the most fundamental of all the properties of radiation—its uniform speed of travel.

A second instance of its failure is provided by an experiment which can be performed in any laboratory. Let S be a source of light, emitting light of approximately pure colour, and let an opaque screen, punctured by two tiny pinholes, A, B, be set up in front of S, so that A and B are at equal distances from S. If we picture the light which S emits as bullet-like photons, then two points P, Q on the laboratory wall will be under fire from S, and we shall expect to find the wall illuminated at the points P, Q and dark everywhere else. In a general way this describes what will usually happen; yet if we make our holes A, B near enough together, the description fails entirely. The most brightly lighted region of the wall will no longer be the points P and Q, but the single point R midway between them, although this is out of both of the lines of fire SAP and SBQ; if light really consisted of bullets, we should expect R to be completely dark. Not only so, but for light of one particular colour, P and Q, which ought to be most brilliantly lighted of all if the light consisted of bullets, may be completely dark.

We can obtain yet more surprising results by blocking and unblocking one of the pinholes, say B. We shall find that so long as B is blocked up, P is brightly illuminated, but the moment B is unblocked, P becomes dark—letting more light in on P changes light into darkness.

The old undulatory theory had provided a perfect explanation of all this; it is, indeed, a special instance of the general principle already explained on p. 122. Let us for the moment think of our diagram as representing the surface of a rectangular piece of water, such as a swimming pool, while A, B is a wall built across the pool with small apertures at A and B. When a swimmer splashes about at S, he will cause ripples to spread over the pond. Some will pass through the apertures A, B, and set up new systems of ripples in the space beyond. These will spread out in circles from the points A, B, and as these two points are symmetrically placed with respect to S, the two sets of ripples will be exactly similar.

Now the point R is symmetrically placed with respect to A and B. Thus the crest of a ripple from A will arrive at R simultaneously with the crest of a ripple from B, and their combined effect at R will be just twice what it would be if only A or B were open separately. On the other hand, P is not symmetrically placed with respect to A and B, so that the two sets of ripples from A and B will not in any case reinforce one another quite so perfectly at P as they did at R. In an extreme case, crests of ripples from A may arrive coincidently with troughs of ripples from B, so that the two will exactly neutralise one another, and the water at P will remain unagitated. If we block up the opening B, the water at P is agitated by the waves which reach it from A. If we now reopen B, and so let more waves in on to P, the water at P becomes quiescent, because we have added a second set of waves which exactly neutralises the first.

These are precisely the results obtained in an actual experiment. They seem nonsensical to the last degree when we picture light as bullets, but perfectly natural and inevitable when we picture it as waves.

Yet suppose we carry our experiment a stage further, and put a sensitised photographic plate against the wall PRQ of our laboratory. If the apertures A and B are both open, this will of course give us a permanent record of the light at R and of the darkness at P and Q. At R the grains of the plate will be changed by the incidence of the two sets of waves, one from A and one from B, which reinforce one another. On the other hand, the grains of the plate at P and Q will undergo no change, because the two sets of waves neutralise one another.

Yet the grains of the plate can only absorb radiation by complete photons, as is shewn by the fact that blue light makes more impression on it than red. Thus, to make our picture consistent, we must suppose that light travels through space in the form of waves, but breaks up into photons as soon as it encounters matter. We shall find later that there is an exactly complementary picture for electrons and protons. This shews us electrons and protons behaving as *particles* while they travel freely through space, and as *waves* when they encounter matter.

There is a complete mathematical theory which shews how in all such cases the particle- and wave-pictures are merely two aspects of the same reality, so that light can appear sometimes as particles and sometimes as waves, but never as both at the same time.* It also explains how the same can be true of electrons and protons. It is hardly possible to give even the vaguest account of this highly intricate theory in non-mathematical terms, but the following considerations will perhaps provide something of a bridge between the particle- and wave-pictures of radiation, and shew how both may represent partial aspects of a unity which transcends both particles and waves.

Free Vibrations

Any system which is capable of vibration is set into vibration when it is disturbed from outside. If the disturbance finally fades away or is withdrawn, the system does not come to rest immediately, but continues to

---

* See, for instance, Heisenberg, *The Physical Principles of the Quantum Theory*, p. 177.

vibrate for a time. The vibrations which it now executes are known as the "free vibrations" of the system, and the periods of these vibrations are described as the "free periods" of the system. The simplest instance of this is provided by a tuning-fork. For all practical purposes, this has only one period of free vibration, which determines the pitch of the note emitted by the fork. Suppose, for instance, this is middle C, which corresponds to 256 vibrations a second. If the fork is disturbed in any way whatever, as by the impact of a blow or the friction of a violin bow, it will be set into vibration, and after the disturbance is over it will be left vibrating at 256 vibrations a second. It will, so to speak, have forgotten what caused it to vibrate, and remember only its own period of free vibration—hence its utility for musical purposes.

A piano string provides a more complicated example of the same thing. When we sound middle C of the piano, the hammer strikes a string which has an infinite number of free vibrations, these being at the rates of 256, 512, 768, 1024, ... vibrations a second. These frequencies are in the ratio $1 : 2 : 3 : 4 : ...$, and the corresponding musical notes are called the "harmonics" of middle C. They are the C above, the G and C above this, then the E, G, B♭, C, D, E and so on in succession, their respective wavelengths being a half, a third, a quarter, and so on, of the length of the string. Again, these tones are associated with the string itself, not with the striking of it. In whatever way the hammer strikes the middle C string, these same harmonic notes are always sounded; only the proportion of their intensities depends on the way in which the hammer strikes the string.

A still more complicated example is the air in a concert hall. This has innumerable free vibrations, their wavelengths ranging from the whole length of the concert hall down to a minute fraction of an inch. When a pianist plays middle C in the concert hall, the hammer momentarily strikes three strings of the piano, and sets them into vibration, after which they are left performing free vibrations of all the wave-lengths and frequencies just mentioned. These vibrations do not persist for ever, because their energy is gradually transferred to the surrounding air through the medium of the sound-board of the piano. We can describe what happens by saying that the piano string sends out waves of sound into the concert hall, but it is an equally fair description to say that the energy of the string is transferred to the free vibrations of the air in the hall; actually it will be transferred almost exclusively to those having the same frequencies of vibration as the string itself, namely middle C and its harmonics. Thus we have two pictures—the wave picture and the free vibration picture—each of which represents the facts equally well, although each represents only one special aspect of the facts.

We have mentioned these cases merely as stepping-stones to a system of still greater complexity, namely the optical laboratory represented in fig. 9 (p. 160). Radiation can travel through this laboratory just as sound can travel through a concert hall, and again this radiation can be represented equally well either as waves or as free vibrations of definite wave-lengths and frequencies. We need not associate the vibrations with any special underlying mechanism, since a theorem of pure mathematics, similar to that already mentioned on p. 62, shews that, quite apart from any special type of mechanism, or indeed of any mechanism at all, every kind of disturbance can be pictured as made up of free vibrations.

If the source of light at S emits radiation of any single definite frequency, the energy it emits is transferred to various free vibrations of the same frequency in the laboratory. If we study these vibrations after the manner of p. 162, we find that, under the special conditions and with the special arrangement of apparatus already postulated, there will be a violent disturbance at R, but no disturbance at all at P or Q. A source of light at S can add to the energy of these vibrations by emitting radiation, and, by the same process reversed, a molecule at R can subtract from their energy by absorbing radiation. Molecules at P and Q cannot, however, do this; we have seen that the vibrations produce no disturbance at P or Q, which shews that there is no coupling between the free vibrations of the laboratory set up by the source of light at S and the molecules at P and Q.

If we now picture the energy of these free vibrations as that of photons, we can say that the source at S emits photons of a definite known frequency, and that molecules at R can absorb these photons, while molecules at P and Q cannot. Thus if we expose a photographic plate on the wall PRQ, the points P and Q will appear dark on the plate, while R will appear light. We can make our picture of the process more vivid by saying that S emits photons which fall on R but not on P and Q. We are now picturing the light as consisting of bullets of energy, but only in a limited sense. We may picture it as bullets when it leaves S, and again when it arrives at R, and this picture will give us a true account of the phenomena actually observed. On the other hand, we must not picture it as bullets while it is passing through the apertures A and B; if we make this mistake we shall expect to find P and Q light and R dark, exactly contrary to the facts of observation. If we want to combine the bullet and wave aspects in a single picture, we must say, as before, that light behaves like waves while travelling through empty space, but like bullets as soon as it encounters matter.

When we adopt the particle picture, we are, in effect, interpreting the energy of free vibrations of any specified frequency as energy of photons of the same frequency, but we must be careful not to identify individual free vibrations with individual photons. The energy of a free vibration in our laboratory extends through the whole of the laboratory, and on imagining the walls of the laboratory to recede to an infinite distance, we find that the energy of a free vibration in space extends through the whole of space. A mathematical theorem shews that the energy of any isolated disturbance in space can be regarded as the sums of the energies of a number of free vibrations, each extending through the whole of the available space. This is true no matter how restricted the area of the disturbance may be, or how large the space may be; inside the area of the disturbance, the different vibrations are cumulative in their effects; outside it, they destroy one another by interference. It is the energies which reside in such restricted areas, not the energies of the separate free vibrations, that must be identified with the photons.

We have not yet found any reason why the energy of

photons should occur only in complete quanta; we shall only understand the atomic aspect of radiation through a study of the properties of matter, to which we now turn.

Atomic Spectra

This problem is most naturally approached through a study of the complex spectra emitted by atoms of the chemical elements. We have seen that striking a piano string in any way whatever causes it to emit sound-waves of various distinct frequencies, which are in the simple ratio $1:2:3:4:\ldots$. For instance, when the fundamental frequency is 256, the other frequencies are 512, 768, 1024, and so on—all the integral multiples of 256 in succession.

In precisely the same way a mass of incandescent hydrogen, or of any other chemical element, emits light-waves of various distinct frequencies, which can be measured with great accuracy in a spectroscope. These frequencies are not found to stand in any such simple ratio as $1:2:3:4:\ldots$; indeed, for a long time, no relation whatever could be discovered between them. Finally minor regularities began to appear. The frequencies of the three most conspicuous lines in the spectrum of hydrogen Hα, Hβ, Hγ were found to be in the ratio $20:27:32$. At first it was conjectured that these vibrations might be the 20th, 27th and 32nd members of a series of harmonics similar to those of a piano string, but it soon became apparent that the vibrations of the hydrogen atom were far more complicated than this. Ritz made a great advance in 1908, when he shewed that all the intricate frequencies of the light emitted by any single substance were connected in a very simple way. He found that there exist a number of fundamental frequencies ν_a, ν_b, ν_c, ... such that the frequencies of the emitted light are the differences between them, namely $\nu_a - \nu_b$, $\nu_b - \nu_c$, $\nu_a - \nu_c$, and so on. Even these fundamental frequencies do not stand in any such simple proportion as the $1 : 2 : 3 : 4 : ...$ of a piano wire, although Balmer and others found that for the hydrogen spectrum they stood in the ratio

$$\frac{1}{1^2} : \frac{1}{2^2} : \frac{1}{3^2} : \frac{1}{4^2} : ...,$$

which is not very much more complicated.

The first step towards the interpretation of spectra must obviously be the assigning of meanings to the frequencies ν_a, ν_b, ν_c, ... etc. This may seem a simple matter, but actually it has presented a problem of very great difficulty. The clue to its solution was found to lie in a suggestion which had originally been made by Bohr on theoretical grounds, and was subsequently confirmed experimentally by Franck and Hertz: *An atom can only exist in certain distinct states possessing different clearly defined amounts of energy. When it passes from one state to another of lower energy, the liberated energy forms a single photon.*

If we know the amounts of energy in these various states, Planck's original quantum law will at once tell us the frequency of the photon emitted at the passage of the atom from one state to another. For if W_a denotes the energy of the atom before the photon was emitted, and W_b the energy afterwards, the amount of energy liberated is $W_a - W_b$, and this must be equal to $h\nu$, where ν is the frequency of the single photon emitted.

The frequency of this photon is accordingly given by

$$\nu = \frac{W_a}{h} - \frac{W_b}{h},$$

and we notice at once that the fundamental frequencies ν_a, ν_b,... of Ritz are merely the energies W_a, W_b, ... of the distinct states postulated by Bohr, all divided by h. Thus the problem of interpreting atomic spectra reduces to that of assigning meanings to W_a, W_b, ..., the values of the energy of the atom in the various distinct states in which it can exist.

When Bohr attacked this problem in 1913, he adopted the then current view that the proton and electron of the hydrogen atom were minute particles charged with electricity, in which case the mutual attraction of the opposite kinds of electricity would cause the electron to describe an orbit round the more massive proton, much as a planet describes its orbit round the sun. In those days the orbit of a planet round the sun was supposed to be determined by the inverse-square law of gravitational attraction, so that it was natural to expect the orbit of the electron to be determined by the similar law of electrical attraction. This law would compel the electron to describe a circle or an ellipse round the proton, but the law of itself gave no definiteness of size to the atom, and so could not limit the atom to distinct states with different clearly defined amounts of energy; it permitted the atom to have any amount of energy.

Bohr accordingly found it necessary to suppose that the orbit not only conformed to this law, *but to certain other laws as well*. These other laws were of the nature of restrictions; they restricted the electron to one or other of a number of definite clearly defined distances from the proton. It was rather as though a number of grooves, some circular and some elliptical in shape, were cut in the space round the proton. The electron had to stay in its groove, but its speed of motion was continually changed as it was accelerated or retarded by the electrical attraction of the proton, just as the speed of a planet's motion is continually changed by the gravitational pull of the sun. Physicists still describe these orbits as Bohr orbits. An electron might go on describing the same Bohr orbit for ever, or it might suddenly fall from one orbit to another of smaller dimensions. When such a fall occurred, the system lost a certain amount of energy. Assuming that this reappeared as a single photon of radiation, Bohr was able to calculate the frequencies of the different kinds of photons which could be emitted, in the way already explained. The frequencies calculated for the hydrogen atom agreed very closely, and probably perfectly, with those actually observed in the light emitted by incandescent hydrogen. This, however, was only a fragment of a far larger problem, and when more complicated spectra of other substances were discussed in the same way, Bohr's theory was found to lead to a less perfect agreement with experiment. In certain cases, it quite obviously gave a wrong result, and no conceivable modification seemed capable of bringing it into line with the facts of observation.

Observables and Unobservables

At this stage Heisenberg introduced a new method of looking at the whole problem, which has proved brilliantly successful. In brief Bohr's theory had pictured an atom as consisting of particles which pushed and pulled one another about in space and time; it represented a last brave but unsuccessful attempt to force nature into a mechanical setting, and to depict the atom as existing in space and time. The difficulties it encountered seem to shew the imperfections of both these concepts. Bohr had pictured the atom as a mechanical structure, but was finally compelled to suppose that at intervals it evaded the limitations of this picture, when it passed from one orbit to another in a wholly non-mechanical way. He had tried to force the electron into space and time, yet was finally compelled to postulate jumps which shewed no continuity in space-time.

Heisenberg did not fetter himself with either mechanical pictures or space-time representations. *A priori*, as we have seen, there are very great odds against our being able to form any kind of visual picture of the fundamental processes of nature. Heisenberg was not prepared to handicap his investigation at the outset by assuming a picture of any kind whatever to be possible. Just as the visual picture of light as waves in an ether had brought confusion into optical theory, so he thought that a picture of an atom as a structure of electrified particles was bound to bring confusion into atomic physics, as indeed it obviously was doing.

In place of Bohr's picture concept of an atom, he introduced a new set of ideas which followed naturally on the changes in scientific outlook which were described in Chapter III, and can perhaps best be approached through a consideration of these changes. These had, in effect, amounted to the dismissal of three concepts from the scheme of science—absolute space, absolute time and the luminiferous ether. Einstein's successes made it clear that these three dismissals had started science on the right road, and by travelling farther along the same road, Heisenberg was now able to bring order into atomic physics. Just as the theory of relativity had removed a whole mass of inconsistencies and contradictions from large-scale physics and astronomy, creating hardly a single new difficulty in their place, so Heisenberg's new line of thought has performed a similar service for atomic physics.

It was not mere accident that selected the three above-mentioned entities for dismissal. Our mental activities are stimulated by sense-impressions which originate beyond our senses; to account for these, we invent an external world of objects and entities, but everything beyond our senses is pure inference. The inferred entities are of many kinds, but fall into two distinct categories, which we may label as "observables" and "unobservables". In brief the distinction is that the observables produce a direct effect on our senses, or on the instruments in the laboratory, whereas the unobservables affect our senses and instruments only indirectly, through the intervention of observables. A typical observable is a photon; a typical unobservable is the ether. Both types of entities may have quantities associated with them, so that we have "observable" quantities and "unobservable" quantities. An observable quantity admits of direct instrumental observation, whereas an unobservable quantity can only be a matter for abstract calculation. Typical of the former is the wave-length of a photon; typical of the latter is the rigidity of the ether.

The universe of the scientist may be expressed diagrammatically somewhat as follows:

Thought
↑
Sense-data
↗ ↖
Instrumental effects
↗
Observables
↑
Unobservables

All the items of the last two categories are purely inferential, but the type of inference is of course different in the two categories. The observables certainly represent something objective, because they affect the senses of everyone in the same way, and affect instruments which are independent of our individual senses, but the very existence of the unobservables is in doubt because they do not affect our senses or instruments at all; unobservables may represent nothing more than bad guesses. In brief, the properties of the observables are inferential, but the very existences of the unobservables are inferential.

We may elaborate our scheme by setting against each category the principal items which belong to it, when it will stand somewhat as follows:

Thought	
Sense-data	Sights, Sounds, Smells, Tastes and Feelings
Instrumental effects	Light, Photographic action, Electric currents, etc.
Observables	Events at hand (impact of photons), Individual space, Individual time
Unobservables	Distant events, Objects, Ether, Absolute space, Absolute time

This represents the universe as it appears to a scientist who explores it with the help of instrumental resources; primitive man would of course short-circuit the item "instrumental effects" and pass directly from observables to sense-data, in the way shewn in our previous diagram.

The observable ingredients of the external world are those which directly affect either our instruments or our senses. At first sight these may seem to be of a vast number of kinds; actually there is only one—the impact of photons. It is obvious that the imprints on photographic plates, which play so large a part in modern experimental science, are the result solely of the impact of photons, and that all optical and photometric effects must be the same. It is less immediately obvious how effects such as galvanometer deflections, which measure the passage of an electric current, or thermometer readings, which measure temperature, or the pressure on an ear-drum which registers the arrival of sound-waves, can be caused by the impact of photons. Yet they are; neither a physical instrument nor a sense-organ can exhibit an effect unless energy is in some way transferred to it, and all energy which is transferred from one object to another consists of photons. We are not of course speaking of photons in the limited sense of bullets of light, but in the more general sense of bullets of energy, which we reach by extending the concept of light to all possible wave-lengths and frequencies. In brief, all instrumental effects and sense-impressions depend on the transfer of energy, and all transfer of energy is by photons. So great a simplification may seem almost too good to be true, but it is not pure gain to the scientist, as we shall find that it imposes very severe restrictions on his exploration of the universe (p. 231).

In addition to the impact of photons, individual space and time may properly figure in our list of observables, as the mental framework in which the arrival of the various photons is set. When we try to arrange and classify the photons which impinge on our senses, we find that the mental creation of a three-dimensional space and a one-dimensional time instantly introduces complete and perfect order.

Out beyond the observables—i.e. still farther away from our senses and instruments—come the unobservables. These are mere pictures, images or models, which science has imagined to exist in reality merely because they seemed capable of inciting the observables to produce the effects actually observed. We have, however, no guarantee that these effects cannot be produced in other ways. To establish that the supposed unobservables exist, it would be necessary to shew that nothing else could produce the observed effects.

The three supposed entities which were dismissed from science in the previous chapter—absolute space, absolute time, and the ether—were all drawn from the list of unobservables, as was of course inevitable. Their fate naturally raises the question of the status of the two unobservables which still remain in our scientific scheme—distant events and objects. Are these also mere bad guesses, made in our hasty efforts to depict, in the light of inadequate knowledge, an external world of which no picture is, in all probability, possible?

It may at first seem surprising that material objects

such as electrons, protons and atoms should figure in our list of unobservables; yet it is here they must be placed. An eventless electron or proton could never disclose its existence to us, and a single electron or proton must necessarily be eventless. The simplest kind of event can affect our senses which needs the juxtaposition of at least two such objects. When the two objects are of the same kind, the event is an encounter of electrons or protons such as can be observed in certain favourable cases. When they are of different kinds, they constitute a hydrogen atom, and the event is the emission (or absorption) of any one of the many types of photons which figure in the hydrogen spectrum. The mere orbital motion of an electron round a proton, which figured so largely in Bohr's theory, is not an observable event. It emitted no light, and so could not affect our senses.

We must then regard electrons and protons merely as unobservable sources of events which are themselves observable. The millions of electrons and protons in the sun exist only as inferences, created to explain the stream of photons which fall on our eyes and skin all day long.

The Stream of Radiation

Heisenberg, holding the views already explained, refused to concern himself with the unobservable electrons and protons in distant atoms, and concentrated on the observable photons which came from them. These form a mixed bag of distinct kinds, which can be distinguished primarily by their frequencies. The bag is less like the bag of animals killed by a sportsman, than like the bag of tickets taken by a ticket collector on a railroad. For the

principle of Ritz shews that each photon has two fundamental frequencies associated with it (its own frequency
being a measure of the distance between the two), just
as each railway ticket has two railway stations associated
with it—e.g. Aberdeen to Birmingham. Bohr's theory had
pictured these two frequencies as those of motions in fixed
orbits, and imagined the emission of a photon to be the
result of the passage of an electron from one orbit to
another, just as we may picture the giving up of a railway
ticket to be the result of a journey from one railway station
to another. Heisenberg did not tie himself to any definite
picture as to the origin of the photons; he was concerned
only with the stream of light. In terms of our analogy,
he did not try to picture in detail the movements of the
trains on the railroad, but studied the workings of the
system as a whole, as put in evidence by the stream of
tickets forwarded to headquarters by the ticket collectors.
For headquarters were our senses, so that the tickets were
observables, but engines, coaches and passengers were all
unobservables.

Headquarters could gain a great variety of statistical
knowledge from such a collection of tickets; they could
discover, for instance, the total number of passengers
leaving Aberdeen, the total number arriving at Aberdeen,
the total money taken at Aberdeen booking-office, the
total number of miles travelled by passengers starting from
Aberdeen, and so on for every station on the railway. If
they wished to co-ordinate all this knowledge, they would
probably begin by tabulating the numbers of tickets issued
from every station to every other station. If we denote
the various stations—Aberdeen, Birmingham, Carlisle,

Dundee, Edinburgh, and so on, by A, B, C, D, E, ..., the knowledge could be put in the form:

$$A \rightarrow B = 23$$
$$B \rightarrow C = 72$$
$$A \rightarrow C = 13, \text{ etc.}$$

It could all be very concisely recorded in a table of double entry, as, for instance:

From	A	B	C	D	E
To					
A	103	23	13	84	22
B	23	207	72	28	43
C	13	72	90		
D	84	28			
E	22	43		and so on	

A square table of double entry provides the obvious way of tabulating quantities, each of which is associated with two others. In mathematics, such a table is called a matrix, and the entries 103, 23, 13, ... are described as the elements of the matrix.

The first entry 103 in our particular table would mean that headquarters received 103 tickets from Aberdeen to Aberdeen; we may if we like interpret these as platform tickets, bought at Aberdeen and finished with at Aberdeen. The entry 23 to the right of this means that 23 people travelled from Aberdeen to Birmingham, while the entry 23 in the line below means that an equal number made the same journey reversed. This may or may not accord with the facts of railway travel, but it does with the original problem of physics. For this is idealised until all conditions become as simple as possible; in particular, the gas which emits the photons is supposed to be in a steady state. This requires that it shall be in equilibrium with its own radiation, and so absorb as many photons of each kind as it emits, except for an insignificant number which escape to affect our instruments. Thus as many atoms pass from one state to another, emitting photons, as pass in the reverse direction, absorbing photons. The result is that in the matrix which specifies the numbers of photons, the corresponding elements are equal, and the matrix is "symmetrical" in the sense that each vertical column contains just the same entries as the corresponding horizontal line.

Our matrix has so far merely specified the numbers—or, perhaps better, the relative proportions—of the different kinds of photons, but this does not contain all the "observable" knowledge at our disposal. A railway ticket has printed on it the names of two places, and also the fare, which gives an indication of the distance between them. In the same way every photon has indelibly and unalterably stamped on it a quantity, its frequency, which tells us the distance between the two fundamental frequencies associated with it.

Even this does not exhaust our observable knowledge. It contains our knowledge of the relative numbers of photons, but not of the manner of their oscillation. A large mass of experimental evidence shews that the oscillations of all photons are like those of a pure musical tone; they can be compared to the up-and-down motion of a point on the rim of a fly-wheel which goes round-and-round with absolute evenness.

In mathematics, such oscillations are described as "simple harmonic". The changes they involve are proportional to those of the quantity $\cos(2\pi\nu t + \epsilon)$, where ν is the frequency of the oscillation, and ϵ fixes its phase (p. 122). Yet such a formula is too explicit for our present problem, in which the phase does not permit of observation.

It is, however, very easy to avoid over-precision of this type. A well-known theorem of Demoivre tells us that, if θ is any quantity whatever,

$$e^{i\theta} = \cos\theta + i\sin\theta,$$

where e is the ordinary exponential function (so that e^x is the quantity e or $2\cdot71828\ldots$ raised to the power of x), $\cos\theta$ and $\sin\theta$ are the ordinary trigonometrical functions, and i stands for the square root of -1. Any algebraic quantity C can be represented in the form $A + iB$, where A and B are ordinary numerical quantities, and Demoivre's theorem now shews that

$$Ce^{i\theta} = \sqrt{A^2 + B^2}\,[\cos(\theta + \epsilon) + i\sin(\theta + \epsilon)],$$

where ϵ depends on the ratio of A to B. If we now assign to θ the specific value $2\pi\nu t$, we find that the first term $\sqrt{A^2 + B^2}\cos(\theta + \epsilon)$ exactly represents the needed oscillation. We may accordingly represent the oscillation by $Ce^{2\pi i\nu t}$, with the understanding that the parts multiplied by i are to be ignored; in using such an expression, we make no claim to know the phase ϵ, because we do not divide C into its constituent parts A and iB. This is a useful and very common artifice in the mathematical theory of oscillations and vibrations of all kinds.

Subject to this understanding, the whole of our observable knowledge of the stream of light received from a mass of, say, hydrogen can be embodied in a single matrix of the form

$$\begin{vmatrix} C_{aa}e^{2\pi i(\nu_a - \nu_a)t}, & C_{ba}e^{2\pi i(\nu_b - \nu_a)t}, \\ C_{ab}e^{2\pi i(\nu_a - \nu_b)t}, & C_{bb}e^{2\pi i(\nu_b - \nu_b)t}, \\ C_{ac}e^{2\pi i(\nu_a - \nu_c)t}, & \text{and so on,} \end{vmatrix}$$

in which ν has been replaced by its known value $\nu_a - \nu_b$, etc., where ν_a, ν_b, ... are the fundamental frequencies of Ritz. We do not know the values of ν_a, ν_b, ... separately; but only their differences, which alone occur in the matrix.

Atomic Structure

Having embodied all our observable knowledge of the light emitted by, say, hydrogen in such a matrix, Heisenberg found the clue to his next step in Bohr's earlier investigation which we have already described. Bohr had pictured the hydrogen atom as consisting of an electron describing an orbit around a proton, and had obtained agreement with observation by supposing that certain "quantum restrictions" only permitted orbits whose diameters were proportional to the squares of the integral numbers. The large outer orbits were nearer together relatively than the small inner orbits, and the largest orbits of all could be regarded as continuous with one another, because their distances apart were insignificant compared with their total dimensions. Here we need no longer think of an electron which passes from one orbit to the next as jumping; we may think of its motion as continuous, and of the change in its energy as a continuous change.

So long as the electron remained in these orbits, Bohr's quantum restrictions were found to have no restrictive effect whatever on its motion, so that Bohr's picture of the atom coincided exactly with the older mechanical picture. As Bohr's picture predicted an emission of radiation which agreed with that actually observed, it follows that the old mechanical picture did the same in this extreme case of atoms of infinite diameter.

In the case of orbits of very large, but not actually infinite, diameter, Bohr shewed that the quantum restrictions only came into play to a minute extent, and that the agreement extended to this case also. This is known as Bohr's "correspondence principle".

Thus the old mechanical picture of nature predicted the radiation from very large atoms with accuracy, although it was known to fail badly for ordinary atoms of small diameter. Heisenberg set himself the problem of modifying the old picture so that it would remain true to observation over its whole range—i.e. for small atoms as well as for large.

As he had resolved to leave unobservables alone, the only material available for the construction of the new picture consisted of the matrix just described, and quantities directly reducible from it. Thus the line of attack was obvious—first to try to reconstruct the old picture, which was valid for very large atoms, in terms of this material, and then to try to extend the picture to cover atoms of all sizes.

Each element in the matrix is of course proportional to the intensity of the corresponding spectral line. If we halve the amount of gas by which the light is emitted, we halve the intensity of each spectral line, and so must halve each element of the matrix. If we reduce each element by dividing it by a suitable factor, we shall obtain a "reduced" matrix which will represent the average light emitted by one single atom.

We have seen how the motion of a particle can always be represented as compounded of a great number of oscillations. In the special case in which the atoms are very large, Heisenberg found that the "reduced" matrix specifies the position of the point electron of Bohr's picture, its separate elements representing the separate oscillations, the compounding of which gives the motion of the electron.

We must not of course expect a similar interpretation when the atom is small. Let us however assume that, even in this case, the hydrogen atom consists of two constituents, which we may still call the proton and electron, without knowing in the least what these words now signify.

As the matrix was found to specify the position of the electron in a large atom, Heisenberg imagined that it must also specify some sort of position of the electron when the atom was small, although, as we have seen, the electron must no longer be compared to a point, but rather to a complete railway system. The elements of the matrix must no longer be taken to represent oscillations which can be merely added together; they rather correspond to the journeys of the various trains on the railway.

It may seem strange that an electron which is describing a large orbit in an atom can be pictured as a particle, while an electron which is describing a small orbit can be pictured as nothing less than a complete railway system. We must, however, remember that Heisenberg's matrix is not concerned with a single photon

originating in a single atom, but rather with a beam of
millions of photons of different kinds originating in
innumerable atoms in different states. For this reason,
it could not be expected to give us a picture of the state
of any single individual atom, but only a composite
photograph of all the atoms. We may if we like say that
the reduced matrix gives us a picture of a "statistical
atom" whose properties and qualities are the average of
the properties and qualities of all the actual atoms concerned in the emission of the light.

Each element in the Heisenberg matrix changes with
the time in a calculable way. On replacing each element
by its rate of change, we obtain a second matrix, which
will give a sort of representation of the changes occurring
in the statistical atom—somewhat like the speeds of
the trains on our railway system; for large atoms, it
simply represents the speed of motion of the particle
electron. If we multiply every element by the mass of
an electron, we obtain a matrix which represents the
momentum of the electron.

Heisenberg's relation

Let us denote this last matrix by p, and the original
reduced matrix, which corresponds to the position of the
electron, by q. Heisenberg found himself able to construct
a new picture of atomic workings in terms of the matrices
p and q, and this is found to agree exactly with observation throughout.

This picture does not involve the matrices p and q
separately, but only their product. If p and q were simple

quantities, such as 6 and 8, the product pq would have an exact meaning, such as 48. With things as they are, the product pq has no meaning as yet, and we are free to assign to it any meaning we like. Actually a mathematical theory of matrices had been in existence long before Heisenberg, and this had assigned a conventional meaning to the product pq, p multiplied by q, and also a conventional meaning to qp, q multiplied by p. If p and q were mere quantities such as 6 and 8, the product pq would of course be the same thing as qp, but when they are matrices, this need not be so. Indeed, on the conventional mathematical interpretations of pq and qp, it is not so; $pq - qp$ has a definite meaning, being itself a matrix which is usually different from zero.

In the extreme case of an atom of very large radius, Dirac has shewn that $pq - qp$ becomes identical with a quantity which occurs in ordinary dynamical theory, and is known as a Poisson-bracket (for our discussion it does not matter what these words mean). Ordinary dynamics tells us that the Poisson-bracket must retain a constant value throughout the whole motion, and we have already seen that ordinary dynamics predicts the radiation from an atom of very large size with accuracy. Thus a true picture of the radiative processes of an atom of very large size can be obtained by supposing that $pq - qp$ retains a constant value.

This picture contains only the "observable" matrix q, and the matrix p which can be directly derived from it. It can obviously be extended to small atoms, since every term in it has a clearly defined meaning for small atoms. The picture is perfectly precise and clear-cut except for the evaluation of the constant quantity on the right of the equation; the only question is whether it is true to observational facts. Now Heisenberg and a great number of other workers in this field have jointly shewn that the whole problem of atomic spectra is solved—not a single old difficulty remaining outstanding, nor a single new one appearing—by supposing that the activities of the electron conform in every case to the simple law

$$pq - qp = \frac{h}{2\pi i}\,\epsilon.$$

In this equation ϵ is itself a matrix, of the kind known as a unit diagonal matrix, namely

$$\begin{vmatrix} 1 & 0 & 0 & 0 \\ 0 & 1 & 0 & 0 \\ 0 & 0 & 1 & \text{and so on} \\ 0 & 0 & & \end{vmatrix}$$

in which all the diagonal terms are equal to unity, while all the rest are equal to zero. Thus the meaning of Heisenberg's equation is that, in the matrix $pq - qp$, all those elements which do not occur in the diagonal are equal to zero (as they would be if p and q were ordinary numbers or quantities), while all the rest are equal to $\frac{h}{2\pi i}$.

Here h is again Planck's constant which pervades the whole of small-scale nature. The letter i stands for $\sqrt{-1}$, which again pervades all atomic physics, possibly as representing a transition from some sort of real time to the time of our observation (p. 100); π stands as usual for the number 3·14159... (the circumference of a circle in terms of its diameter), while—somewhat refreshingly, after all the other symbols—2 merely denotes the ordinary 2 of arithmetic, twice one.

The central fact in the situation is that pq is not equal to qp. There is nothing mystical or surprising in this, since p and q are not single quantities, but arrays of quantities, and the product pq is defined in such a way as to make it different from qp at the very outset. Yet the quantities which enter into the arrays p and q are interlocked in a curious way, so that p and q may almost be treated as single quantities rather than as arrays of independent quantities. This appears very clearly in the limiting case of atoms of very large dimensions, in which, as we have seen, we may think of p and q as the co-ordinate and momentum of a point-electron.

A simple illustration may explain the nature of this interlocking. Any pair of independent numbers, say 2, 3, forms an array of two numbers, but out of these we may form the algebraic combination $2 + 3 \sqrt{-1}$, which may be treated as a single number—it is for instance a root of the quadratic equation $x^2 - 4x + 13 = 0$—although its constituents 2, 3 do not lose their identities. If we take P and Q to be the numbers whose constituents are 2, 3 and 4, 5 respectively, the product PQ has the value

$$PQ = (2 + 3 \sqrt{-1})(4 + 5 \sqrt{-1}) = -7 + 22 \sqrt{-1},$$

which is of the same nature $(-7, 22)$ as the original numbers.

Here of course the product PQ appears to be precisely equal to QP. Yet there is an ambiguity here, since we have taken no account of the fact that there are two distinct square roots of -1, of equal values but of opposite signs.

Let us denote these by i and j, and further agree that any product PQ is to be obtained by putting $\sqrt{-1}$ equal to i in the first factor, and equal to j in the second. With this definition of a product, PQ is not the same thing as QP; in fact we find that

$$PQ - QP = (2+3i)(4+5j) - (4+5i)(2+3j)$$

of which the value is $2(i-j)$ or $4i$.

In some such way as this, we may treat p, q almost, but not quite, as single quantities, q being analogous to the co-ordinate and p to the momentum of a point-electron. To the extent to which this analogy holds, we find that pq is analogous to the action (p. 124) of the electron throughout the description of a complete orbit* measured in one way, while qp is analogous to this same action measured in a different way. We now see that the equation of Heisenberg,

$$pq - qp = \frac{h}{2\pi i},$$

merely expresses a relation between two different quantities, each of which becomes analogous to the action in an electron-orbit in the limiting case in which this orbit is of very large dimensions.

Although this equation has so far only been found true for an "average" or statistical atom (p. 185), yet, as so often happens in science, its validity appears to extend far beyond the special conditions which led to its discovery. The state of any dynamical system can be specified by a number of co-ordinates (q), and its motion by the values of the momenta (p) associated with these co-ordinates. With any such pair of corresponding values assigned to p and q,

* More precisely, $2\pi pq$ and $2\pi qp$ are analogous.

Heisenberg's relation appears to be confirmed by observation, although we must remember that the observations can never be made on single atoms, but only on statistical assemblies.

This generalisation implies that the equation must hold for individual atoms, as well as for statistical atoms. This is necessary to explain why a mass of gas gives a spectrum of sharp lines. It also shews that each individual electron, when *inside* an atom, has the complexity of a whole railway system rather than of a simple moving point; the picture of an electron as a point in space and time fails, completely and finally.

To take a further instance, the atoms of a solid body are in effect a vast number of oscillators vibrating around positions of equilibrium, the energy of their oscillations being merely the heat-energy of the solid. Heisenberg's relation shews that such an oscillator can never be completely devoid of energy, its oscillations necessarily possessing $\frac{1}{2}$ or $\frac{3}{2}$ or $\frac{5}{2}$ or any such number $(n+\frac{1}{2})$ of quanta of energy $h\nu$. Thus a solid body can never lose all its heat; even at the absolute zero of temperature, each of its internal oscillations possesses half a quantum of energy. Surprising though this conclusion may be, it is strikingly confirmed by investigations on the specific heats of solids at low temperatures.

The relation can also be applied to the rotations of atoms and molecules, when it shews that the angular momentum must always be one or the other of certain definite and calculable multiples of h. Transitions from one of these values of the angular momentum to another produce the lines observed in band spectra, and the calculated frequencies are confirmed by observation.

Perhaps, however, the most interesting application of the equation is to ordinary radiation in empty space. We have seen (p. 167) how the energy of such radiation can be regarded as made up of the energies of a number of separate free vibrations. Heisenberg's relation shews that the total energy of the free vibrations of any given frequency ν must be an integral multiple of $h\nu$, which is precisely the supposition on which Planck originally based his quantum theory of radiation. In other words, the relation shews that all radiation can be regarded as consisting of indivisible photons each having energy $h\nu$; it brings atomicity into the wave picture of radiation, and this, in combination with the considerations of p. 165, establishes the equivalence of the wave picture and the particle picture.

All this suggests that Heisenberg's relation must be the expression of some quite fundamental law which, so far as we can at present see, must hold throughout the whole of nature. We shall discuss possible interpretations of this law below (p. 209).

Transition to Newtonian Mechanics

The quantities of action which occur in the large-scale events of everyday life are of course enormously large multiples of h. For instance, the action for a complete oscillation of the pendulum of a grandfather clock is about

400,000 seconds × ergs,

while the value of Planck's constant is only

$h = $ 0·000000,000000,000000,000000,006555 seconds × ergs.

The products pq and qp which occur in Heisenberg's equation are each of the nature of action, and for tangible bodies they will be such enormous multiples of h, that the single h on the right-hand side may be neglected in comparison. When this is done, the equation takes the form

$$pq = qp,$$

and merely tells us what was accepted without hesitation for the whole of nature, until challenged by Heisenberg.

This shews very clearly how the new theory of Heisenberg gradually fades into the ordinary mechanical theory of Galileo and Newton as we pass from atomic structures to objects of tangible size. Primitive man did not regard a river as a collection of molecules of water but as a continuous stream, and his more sophisticated descendants, still treating water as continuous, developed the science of hydraulics. This was only suited for dealing with molecules in vast crowds; it gives accurate results for rivers in which billions of molecules are involved, but fails for single molecules. In the same way the Newtonian mechanics was only suited for dealing with processes in which the unit of action h occurred in vast crowds; it gave accurate results for the motion in large-scale processes, where billions of units of action were involved, but failed for sub-atomic processes which involved only single units. Newtonian mechanics is the limit to which Heisenberg's picture of nature proceeds when the units of action become so numerous that they can be treated as a crowd.

We may also compare Heisenberg's equation with the earlier quantum restrictions of Bohr, of which it is of course the more accurate successor. It used to be supposed that these restrictions were somehow added to the

Newtonian laws; large systems were subject only to the Newtonian laws, while atomic systems were subject to these laws and to the quantum restrictions *in addition*. Heisenberg's equation is occasionally discussed in a similar manner, as though the laws of nature allowed less liberty to little systems than to big. Actually this is not the case. The information that little systems obey Heisenberg's law only corresponds to what we have already tacitly assumed about the big systems in supposing that for them pq is the same thing as qp. We have perfectly unwittingly assumed this in supposing that big objects can be represented in time and space. The fact that pq is not the same thing as qp for sub-atomic nature casts doubt on whether sub-atomic nature can be represented in time and space at all, a question to which we shall return later (p. 253).

Somewhere in Heisenberg's equation the innermost nature of atomic structure must lie hidden, if we could but read the riddle aright. As the equation does not bear any obvious interpretation on its face, our best procedure will be to try to construct a kind of model system which shall conform to the laws expressed in the equation. If our attempt succeeds, the model will not necessarily, or even probably, be identical with any real structure in nature, but is likely nevertheless to throw some light on the nature of the atom, for it would be surprising if two distinct systems, both governed by the same equation, did not have some properties or characteristics in common.

CHAPTER VI

WAVE-MECHANICS

While Professor Heisenberg of Leipzig was following up the train of thought described in the preceding chapter, Prince Louis de Broglie of Paris and Professor Schrödinger of Berlin were engaged on an independent attack on the problem of the structure of matter. Between them, they devised an alternative explanation of the origin of the spectra of chemical substances, which at first sight seems to bear but little relation to that of Heisenberg. Subsequently Schrödinger himself, as well as Born and Wiener, shewed that the two sets of ideas not only led to the same results, namely those actually observed in nature, but were fundamentally identical. Schrödinger had in effect obtained a solution of Heisenberg's equation which admitted of physical representation, and so provided us with a sort of model of the electron. This model has proved capable of interpreting all the results actually observed by spectroscopists. We shall first explain the work of de Broglie and Schrödinger from this standpoint, although it will be understood that in the first instance its authors achieved it entirely independently of Heisenberg's ideas. Later we shall discuss the more physical concepts of de Broglie and Schrödinger in a less mathematical manner; some readers may prefer to pass directly to this discussion (p. 204).

The Wave Picture

If any quantity r increases n times as rapidly as a second quantity q, then we describe n as "the differential coefficient of r with respect to q", which we write as $\frac{d}{dq}(r)$.

Any other quantity, such as the product qr, will also have a differential co-efficient with respect to q, which we write as $\dfrac{d}{dq}(qr)$.

The product qr has a double cause of change, namely the changes in q and the changes in r. On account of the first cause, qr changes r times as fast as q; on account of the second, q times as fast as r. Adding these changes together, we obtain the relation

$$\frac{d}{dq}(qr) = r + q\frac{d}{dq}(r).$$

We notice that r enters just once in each term of this equation, so that it can be written in the alternative form

$$\left(\frac{d}{dq}q - q\frac{d}{dq} - 1\right)r = 0,$$

in which we regard $\dfrac{d}{dq}$ as an operator, this meaning that everything which comes after it must undergo differentiation with respect to q. We also regard q itself as an operator, meaning that everything that comes after it is to be multiplied by q, this latter being of course the meaning in ordinary algebra. As our equation is true whatever r is, we may write our knowledge in the form

$$\frac{d}{dq}q - q\frac{d}{dq} = 1,$$

in which every symbol is treated as an operator, including unity on the right-hand side.

If we think of a number, then double it and then halve it, the final result will of course be the number we first thought of. In the same way, we may think of any

quantity, perform upon it all the operations directed by
the combination on the left-hand side of the above equation, and our equation shews that the result will always be
the quantity we first thought of. If we multiply both sides
of the equation by $\dfrac{h}{2\pi i}$, it reads

$$\frac{h}{2\pi i}\frac{d}{dq}q - q\frac{h}{2\pi i}\frac{d}{dq} = \frac{h}{2\pi i},$$

which, we see at once, is of the same general form as
Heisenberg's relation. Indeed, the two relations become
identical if we take

$$p = \frac{h}{2\pi i}\frac{d}{dq}.$$

When we say that $x = 3$ is a solution of $x^2 + 2 = 11$, we
mean that the substitution of 3 for x reduces the equation
to a truism. In the same sense the value we have just
written down for p reduces Heisenberg's equation to a
truism, so that in a certain sense we may say that it is a
solution of this equation. We must not say it is *the* solution,
or the only solution, any more than that $x = 3$ is the only
solution of $x^2 + 2 = 11$, but it is *one* solution, and so will
shew us something of the meaning of the equation. In
more physical language, it shews us how to construct a
model or picture, which may be only one of many possible
models or pictures, and yet may perhaps tell us something of the physical meaning of the equation.

Let us first use it to examine what Heisenberg's equation
means when applied to the motion of a particle, such as
an electron. We need not suppose that the particle is so
minute as to be point-like, but we shall suppose that its
position can be specified by the position of a point in

space, just as the position of a cricket ball can be specified by the position of its centre, or that of a train by the foot-plate of its engine. Then we can specify its position (as on p. 86) by an array of three co-ordinates x, y, z, which we may regard as giving the distances through which the particle has moved in three directions at right angles to one another—as, for instance, vertical, north-south and east-west. The speed of motion of the particle will be specified by a similar array u, v, w, and its momentum by yet a third array a, b, c, in which, however, we know that a is the same thing as mu, where m is the mass of the particle. In this simple case, the "solution" we have already obtained for Heisenberg's equation assumes the form

$$mu = \frac{h}{2\pi i} \frac{d}{dx} \qquad \ldots\ldots(C),$$

and there will of course be similar values for mv and mw, the system of three equations thus being symmetrical in the three co-ordinates x, y, z. We have, however, seen (p. 100) that any true picture of nature must be symmetrical in the four co-ordinates x, y, z and ict of the continuum. This indicates that our system of three equations is incomplete; there must be a fourth equation corresponding to the co-ordinate ict. This is easily found to be

$$mc^2 = -\frac{h}{2\pi i} \frac{d}{dt} \qquad \ldots\ldots(D).$$

If these equations seem meaningless, this is only what was to be expected. Heisenberg began by conjecturing that it would prove impossible to construct an intelligible picture or model of an electron. The fact that our equation—the first step towards such a model—already proves to be unintelligible, suggests that he may have been right. If the equations had led us to a simple model, such as a tiny hard sphere, Heisenberg would have stood convicted of unwarranted pessimism.

We can, nevertheless, infuse a little more physical meaning into our equations by remembering that both sides are operators, and so are hungry for something on which to operate. If each member of the first equation (C) is given a symbol ψ on which to operate, it becomes

$$\frac{d\psi}{dx} = \frac{2\pi i\, mu}{h}\, \psi,$$

which is a very familiar equation. Solving it, we find that ψ must be of the form

$$\psi = Ce^{2\pi i\left(\frac{mu}{h}\right)x},$$

where C is a constant.

The formulae given on p. 181 shew that this represents a train of regular waves. As x changes, the value of ψ fluctuates between the values $+C$ and $-C$, repeating itself at regular intervals h/mu. It is beyond the scope of this book to enter upon either a rigorous or a comprehensive discussion of mathematical formulae. The present discussion is neither, but shews that a momentum mu is *in some way* related to a train of regular waves of wave-length h/mu, or h divided by the momentum in question.

Photons have already provided an instance of this, for we have seen that the wave-length of a photon is equal to h divided by its momentum. There is experimental evidence that the relation is equally true for electrons.

In the experiments of G. P. Thomson, a shower of electrons, all moving with the same speed and in the same direction, like a regiment of soldiers marching in perfect order, was allowed to strike on a very thin film of metal. An older physics would have predicted that each electron would fight its individual way, as best it could, through the atoms of the film and their interstices, so that the current of electricity would emerge on the other side as a disordered mob of electrons, the individuals moving at different speeds and in different directions. Instead of this, the experiments shewed that they form a perfectly regular wave pattern such as is shewn in fig. 2 of the frontispiece. Their scramble through the interstices of the solid has not introduced disorder into their formation but a new kind of order. It has changed the quality of what order there was, and given it wave-like characteristics.

It is found possible to measure the precise wave-lengths of these waves. Their wave pattern proves to be identical with that which is formed by X-rays of a certain known wave-length, so that this must be the wave-length of the shower of electrons also. Now the wave-lengths obtained in this way are found to obey the theoretical law exactly—invariably the wave-length is h divided by the momentum of each electron in the shower. It is important to understand that this wave-length has nothing to do with the spacing of the atoms in the metal film.

Experiments on showers of positively charged protons have given similar results.

Thus we may proceed with some confidence that we are on the right road; our picture, however unintelligible it may be to us, is true to reality.

The wave-length of the electron waves is determined by

the speed of travel of the shower of electrons, and this shews that the waves cannot have any objective existence when the shower is travelling through empty space. For under these conditions "speed of travel" has no meaning at all—this of course is the main message of the theory of relativity. As soon as a shower of particles encounters matter of any kind or an electric field, this provides a frame of reference against which the speed of motion can be measured, and the expressions "speed of motion" and "momentum of a particle" acquire a definite meaning. There is no reason why the waves should not be real now, but we shall soon see that even in this case they must not be supposed to possess any material or substantial existence.

We can discuss the fourth equation of our group, equation (D) on p. 197, in precisely the same way. If we provide a quantity ψ for each side to operate on, we get an equation of the same type as before, of which the solution is found to be

$$\psi = Ce^{-2\pi i \left(\frac{mc^2}{h}\right)t}.$$

This represents vibrations which repeat themselves at regular intervals h/mc^2 of time. Again we cannot enter on a rigorous or comprehensive mathematical discussion, but we see that the existence of a particle of mass is *in some way* associated with vibrations of frequency mc^2/h.

The theory of relativity tells us that a particle of mass m is a storehouse of energy of amount mc^2. Thus the frequency of vibration is equal to the energy divided by h.

Again photons provide an instance of this, their frequency being equal to their energy divided by h.

The theory of relativity shews that the mass of a moving

particle depends on the speed of motion of the particle, being proportional to

$$\frac{1}{\sqrt{1 - \dfrac{u^2}{c^2}}},$$

the factor we have already encountered in Chapter III (p. 89). Thus a moving particle has a greater mass than one at rest, and so has also greater energy; actually the excess is precisely the kinetic energy of its motion. It follows that a moving particle is associated with more rapid vibrations than one at rest; its motion increases the frequency of vibration.

It might seem at first that these vibrations must be subjective, like the waves just discussed, because they depend on the speed of motion of the particle, and we have no objective framework against which to estimate this speed. Actually this is not the case. Motion with a speed u increases the rate of oscillation by the factor mentioned above, but also, as we saw on p. 91, it increases the unit of time in precisely the same ratio, so that the two effects cancel out. Thus no reason has so far appeared why these vibrations should not be real. If we picture the electron as oscillating mc^2/h times a second, the wave-pattern of an electron is very simply explained as resulting from the interaction of this oscillating system with the solid surface on which it falls. Calculation shews that the wave-lengths disclosed by actual experiment would be produced by the electron-structure vibrating at the rate of $1 \cdot 24 \times 10^{20}$ (124 million million million) complete oscillations a second. Thus we have a far more wonderful and surprising picture of the electron than that which exhibits it as a tiny billiard ball charged with electricity, although it must be remembered that we are still in the dark as to how far these vibrations are real, and how far, like the waves, they are mere mathematical fictions. Theory and observation agree in suggesting that protons may also be pictured as performing vibrations, real or fictitious, at the even greater rate of 229,000 million million million complete oscillations a second.

We have seen that both waves in space and vibrations in time are associated with the motion of a particle. We can combine the two effects in the single formula

$$\psi = Ce^{2\pi i \left(\frac{mu}{h}\right)\left(x - \frac{c^2}{u}t\right)},$$

this value of ψ satisfying equations (C) and (D) simultaneously.

It represents waves of wave-length h/mu, travelling in the direction of x at a speed c^2/u. This speed again is one with which the theory of relativity has made us familiar; it is the speed of propagation of local time (p. 91). Thus we see that a particle of mass m moving at a speed u is *in some way* related to such a train of waves.

A motion which is performed with varying speed is related to a more complicated system of waves. As an illustration, let us consider the special case of an electron moving in a field of electric force. To simplify the problem we shall suppose that its speed always remains small in comparison with that of light.

As we have seen, its motion must conform to the laws of ordinary mechanics as well as to Heisenberg's restriction. Ordinary mechanics tells us that a moving electron possesses kinetic energy of amount $\frac{1}{2}m(u^2+v^2+w^2)$; and that it moves in such a way that the sum of this and its potential energy, which we may call V, retains a constant value throughout the motion. If we denote this constant value by E, we know from ordinary mechanics that

$$\tfrac{1}{2}m\,(u^2+v^2+w^2) = E - V \qquad \ldots\ldots\text{(E)}$$

throughout the motion, while Heisenberg's relation adds the further restrictions which we can represent in our model by replacing mu by $\dfrac{h}{2\pi i}\dfrac{d}{dx}$, and so on. The combined information leads to the equation

$$-\frac{h^2}{8\pi^2 m}\left(\frac{d^2}{dx^2}+\frac{d^2}{dy^2}+\frac{d^2}{dz^2}\right) = E - V.$$

In this equation, as before, every term is to be regarded as an operator. Let us again supply a symbol ψ for the operators to act on, without for the moment pausing to enquire what ψ is. Our equation now becomes

$$\left(\frac{d^2}{dx^2}+\frac{d^2}{dy^2}+\frac{d^2}{dz^2}\right)\psi + \frac{8\pi^2 m}{h^2}(E-V)\,\psi = 0 \quad \ldots\text{(F)}.$$

The value of ψ will still oscillate with the time, so that mathematicians will recognise the equation as expressing the propagation of waves. The wave-length is no longer definite; it varies from point to point, being proportional to $1/\sqrt{E-V}$, just as the wave-length of light varies as it passes through a refracting substance. We see that an electron moving in an electric field is *in some way* related to waves like those of light moving through a refracting substance.

De Broglie Waves

The equation we have just obtained is generally known as Schrödinger's wave equation, because Schrödinger first obtained it by a distinct method of his own. This method did not involve passing through the particle concept of an electron at all. Like Heisenberg, Schrödinger was convinced that Bohr's earlier theory had failed because it envisaged the electron and proton in the too precise and too concrete form of small charged particles. Before Schrödinger came on the scene at all, this failure had reminded de Broglie of the earlier, and very similar, failure of the corpuscular theory of light. A theory which had pictured light as a shower of minute corpuscles had explained shadows and other simple large-scale properties of light, but only a wave picture had been found capable of explaining its more subtle small-scale properties. In the same way the picture of matter as a collection of minute particles, namely electrons and protons, explained some but not all of its properties, and these were mainly the large-scale properties. De Broglie suspected that a wave picture might be needed to explain the remainder.

Whatever an electron may be, it must be supposed to conform, like the rest of nature, to the theory of relativity. This shews that it is meaningless to assign to an electron properties which can be specified in space alone; their description must involve time as well. This may seem a small clue to go on, but in actual fact it is found to restrict the structure and behaviour of the unknown object somewhat drastically. We find that if the electron conforms to the theory of relativity, it must be possible to picture its structure mathematically as a system of waves.

For if the structure of an electron at rest is specified in terms of x, y, z and t, then when the same electron is moving with a speed u in the direction of x, its specification will be the same except that x must be replaced by $x - ut$ and t must be replaced by the local time $t - \dfrac{xu}{c^2}$. This is of course the content of the Lorentz transformation, except that we have disregarded certain small changes which only become important when the electron is moving with a speed comparable with that of light.

The change of x into $x - ut$ is a necessary consequence of the motion of the electron; $x - ut$ retains its value if the electron moves with a speed u in the direction of x. But the change of t into $t - \dfrac{xu}{c^2}$ implies motion of a different type; for this quantity to retain its value, an unknown something must move with a speed c^2/u in the direction of x. We may picture this as the propagation of some kind of disturbance, or a system of waves, moving in the same direction as the electron at a speed c^2/u, the speed at which local time is propagated.

A group of waves, each of which was travelling at precisely the same speed, would of course itself travel at the same speed as its individual waves. A flash of light provides an obvious instance of this: it travels at the same speed as the waves of light which compose it. But clearly we cannot imagine an electron represented by such a group, since waves travelling at a uniform speed c^2/u would immediately run away from the electron, which only travels at the slower speed u.

Let us now consider the motion of a group of waves,

anywhere and of any kind, which travel at different speeds, but momentarily extend through only a small region of space—a storm at sea will help to fix our ideas. The front of the disturbance will of course forge ahead at the speed of its fastest wave, but its tail will travel only at the speed of the slowest. Thus the head and tail will continually increase their distance apart, so that the group will spread, while the centre of the group, always lying between the two, will move at a speed intermediate between those of the fastest and slowest waves.

There is one case in which these conclusions do not follow, namely when each constituent wave is already spread through the whole of space. This is not incompatible with the whole system forming a group of finite, or even small extent, because the outlying waves may neutralise one another completely by destructive interference (p. 167). Indeed the mathematical theorem to which we have already referred has shewn that any disturbance, no matter how restricted, can always be regarded as made up of constituent waves, each of which extends through space. A group of waves which is restricted to a small extent of space in this way is known as a "wave packet".

Analysis shews that a wave packet will in general spread, but may do so only slowly. When it spreads so slowly as to remain a compact structure through an appreciable time, we may properly speak of its speed of motion, and we find that this need no longer be intermediate between the speeds of the fastest and slowest waves which enter into its composition. For instance, a wave packet may itself travel with a speed u, although each of its constituent waves may have a speed of travel near to some quite different speed, such as c^2/u. This is of course exactly what we want if we are to explain the electron as a group of waves. The condition that the speeds of the group and of its individual waves shall be related in this particular way is quite simple; it is that the frequency of a wave travelling at a speed c^2/u shall be proportional to

$$\frac{1}{\sqrt{1 - \dfrac{u^2}{c^2}}},$$

which is exactly the relation we have already obtained for the waves we discussed on p. 201.

Thus we can identify the de Broglie waves we are now discussing with those previously discussed, the wave-length and period of vibration being given by the formulae already mentioned, and so proving to be the same for electrons, protons and photons.

A detailed study confirms that this system of waves will not run away from the electron. Individual waves are continually neutralising one another before and behind, and reinforcing one another in the intermediate regions, and they do this in just such a way as to build up a permanent structure which moves with precisely the speed of the electron. Except for the fact that the electron-waves are purely mathematical, their action proves to be similar to what we see in the ordinary bow wave of a boat. Although ripples on the surface of the sea may travel faster or slower than the boat, this complete system of ripples does not run away from the boat, but progresses steadily at exactly the speed of the boat.

The speed of light c is necessarily greater than u, the speed of motion of the electron. The speed c^2/u of the waves is greater than the speed of light in just the same proportion, so that the waves not only travel faster than the electron, but faster than light itself. For instance, if the electron is travelling at a quarter the speed of light, the individual waves travel at four times the speed of light; if the electron is a slow one, travelling at only a thousandth part of the speed of light, its waves travel a thousand times as fast as light, and so a million times as fast as the electron.

This of course only gives us a mathematical picture of a kind with which mathematicians are very familiar; they are accustomed to regarding all kinds of changes as produced by successions of waves, merely as a convenient means of description.

We have seen that the wave-lengths and periods of electrons, protons and photons are all given by the same formulae, namely

$$\text{momentum} \times \text{wave-length} = h,$$

$$\text{energy} \times \text{period} \qquad = h.$$

It is certainly very remarkable that these formulae are the same for three such dissimilar objects as electrons, protons and photons. One possible explanation is of course that these apparently diverse objects may in the last resort be of the same nature, at any rate so far as their oscillations, as expressed by their frequencies and wave-lengths, are concerned. The fundamental distinction that electrons and protons carry electric charges, whereas photons do not, is enough in itself to account for many of the differences of their properties. Both carry energy and so possess mass. For a charged particle to carry a finite amount of energy, it must move more slowly than light, whereas for a photon to carry a finite amount of energy, it must move at precisely the speed of light. This explains why photons always travel with the speed of light, while electrons and protons travel more slowly. Again, electrons and protons interact with one another through the attractions and repulsions of their electric charges, whereas photons, having no charges, cannot interact with one another at all. Thus there is considerable justification for regarding photons as being of the same nature as electrons and protons, but without electric charges.

Yet the facts admit of a far wider interpretation than this. The formulae in question are direct consequences of Heisenberg's relation, and the whole of the available observational evidence indicates that this relation is true throughout all nature; its validity certainly extends beyond electrons, photons and protons (p. 189).

Hence it seems probable that the derived formulae also are valid throughout all nature, the cases of photons, electrons and protons only providing illustrations of a quite general truth. In other words, the formulae may perhaps express some general property of space and time, rather than of special phenomena or objects in space and time. Energy for instance may be merely another way of regarding frequency or time, and momentum only another way of regarding wave-length or space. Just as we can regard light almost indifferently as either waves or particles, so it may be that we can regard space almost indifferently as either extension or momentum, and time as either oscillations or energy.

There may seem to be two distinct conjectures here; actually there is only one. For the theory of relativity shews that space is related to momentum in the same way in which time is related to energy. When the imaginary observer of the theory of relativity changes his speed of motion through space, he changes space into time in the sense already explained; and in precisely the same sense, and to the same extent, he changes momentum into energy. In the space-time continuum, momentum and energy are merged into one, like space and time themselves.

Thus the conservation of energy may admit of interpretation also as a conservation of oscillations; the total number of oscillations taking place throughout the universe in unit time may remain constant, and this may give an absolute measure of time. Similarly the conservation of momentum may admit of interpretation as a conservation of wave-number (number of waves per unit of length), and this may give an absolute scale of length.

Modern science has taken a great deal away from the nature of nineteenth-century science, and may be expected to supply something to fill the gaps it has created. The concept of relativity seems to qualify as one of the new factors; possibly the concept just mentioned may be another. The mere circumstance that we have to consider such possibilities seriously shews how far removed the science of to-day is from that of thirty years ago; if, as we hope, present day science is one stage on from the world of appearance, as represented by the older science and the "common-sense" view of nature, towards the world of reality, the same circumstance may also suggest how wide is the gulf between appearance and reality.

In any event, the frequencies of electrons and protons ought to possess the same kind of reality as the corresponding quantities for photons. Now we know something, at least, as to the reality of these latter frequencies.

If we turn a dynamo at such a rate that a coil passes through a magnetic field 50 times every second, we shall obtain an alternating current with a frequency of 50 cycles a second. Through whatever magnetic leakage there may be, electric waves will travel out into the surrounding space, and these waves will also have a frequency of 50 cycles a second. If we picture the radiation as consisting of photons, then the frequency of the photons will still be 50, and this frequency must be just as real in time as that of the turbines which drive the dynamo.

It is much the same with wave-length of photons. When electric waves are sufficiently long, we can map them out with a spark-gap. We may find, for instance, that we have to walk 50 feet to pass from one crest to the next. If we picture the radiation as photons, we must suppose their wave-length to be 50 feet, and this wave-length would seem to be just as real a length in space as the 50 feet we have to walk in order to measure it.

This may seem to suggest that the wave-lengths of photons are real, although we have just seen that the wave-lengths of electrons cannot be. We shall soon see that there is no real contradiction; a want of reality pervades all and everything, creeping in from a quite unexpected direction—a direction, at any rate, which must seem very surprising to a mind brought up to think in terms of the objective concepts of the older physics.

The Nature of Electron Waves

We have obtained a complete mathematical specification of electron waves, but this tells us nothing as to the true nature of the waves themselves. Out of the waves which we find to be connected with the electrons and protons of the atom, mathematical theory can reconstruct the wave-lengths of the photons which the atom ought to emit, and finds a perfect agreement with observation. This tells us something about the wave-lengths, but tells us nothing about the waves except their lengths. We have little but conjecture to help us discover of what the waves actually consist.

It was at first conjectured that they consisted of electricity. The most directly observable property of the electron is that it carries a charge of electricity of un-varying amount—it seems to be deprived of most of its other supposed properties, such as minuteness, hardness, sphericity, by the wave-mechanics. Thus when it was first suspected that the electron had a structure, it was natural to think of this as a structure of electricity. Yet there are two distinct reasons why this concept cannot be maintained.

In the first place, it is a universal property of every kind of wave to scatter through space. We may, for instance, picture a proton at any one instant as a packet of waves, occupying a diameter of a hundred millionth part of a centimetre, which is roughly the diameter of the hydrogen atom, but the waves will rapidly spread so as to occupy more space than this. Ehrenfest has calculated that such a bundle of waves would double its linear dimensions in a ten million millionth part of a second, so that obviously such a system of waves would soon grow too big to shew the spatial properties of a proton. A smaller bundle of waves would expand even more rapidly.

Mathematical theory shews that it is quite impossible to devise a system of waves which shall not scatter at all. Suppose, however, that we could devise a system of waves which would not scatter to any appreciable extent, while a proton or electron was pursuing an undisturbed path through empty space. Even so, the waves must scatter as soon as the particle interacts with other matter; we have direct experimental evidence of this in the wave patterns they form on a photographic plate. Thus, if the wave structure of a proton or an electron merely represented its electric charge, this would be scattered as soon as the particle encountered matter. Yet observation shews that this does not occur; electrons and protons maintain their identity, and preserve their charges intact.

The second objection is even more directly fatal. Let us consider what will happen to the waves of one electron when it meets another electron, and the two exert electric forces on one another. Again the motion of each electron must conform both to the old mechanics and to Heisenberg's conditions. The old mechanics tells us that the total energy, which consists of the sum of the kinetic energies of the two moving particles with the potential energy added on, retains a constant value which we may again denote by E. This is expressed by the equation

$$\tfrac{1}{2}m\left(u^2+v^2+w^2\right) + \tfrac{1}{2}m'\left(u'^2+v'^2+w'^2\right) = E - V,$$

where the accented symbols m', u', v', w' refer to the second particle. We can now represent Heisenberg's further conditions by replacing $m'u'$ by $\dfrac{h}{2\pi i}\dfrac{\partial}{\partial x'}$, etc., and we obtain, in place of our previous wave-equation (F) of p. 203, the new equation

$$\left(\frac{d^2}{dx^2}+\frac{d^2}{dy^2}+\frac{d^2}{dz^2}+\frac{d^2}{dx'^2}+\frac{d^2}{dy'^2}+\frac{d^2}{dz'^2}\right)\psi+\frac{8\pi^2 m}{h^2}(E-V)\psi=0$$

$$\dots\dots(G).$$

This equation still represents the propagation of waves, but no longer in the ordinary space of three dimensions, having x, y, z as co-ordinates. The waves are propagated in a six-dimensional space having x, y, z, x', y', z' for co-ordinates. In the same way, if a million electrons met, their waves would be propagated in a space of three million dimensions. Such a space can only be regarded as a mathematical fiction, and as we cannot suppose waves to be more real than the space through which they are propagated, the waves must be of the same nature.

We have seen that the waves of a single electron are propagated in a space of only three-dimensions. We might be tempted to identify this with the space of everyday life, and conclude that the waves were real, were it not for the possibility of this electron meeting a second electron. When two electrons meet, they meet on an equal footing, so that neither of their sets of waves can claim a greater reality than the other. If we are asked to say which set of waves is real, we can only perform a judgment of Solomon, and declare that both are fictitious. And this makes it impossible that the waves should consist of electricity, or indeed of anything else which exists in our ordinary everyday space.

Yet before leaving this as a final statement of truth, we

shall do well to consider precisely what we mean by "fictitious".

Photon Waves

The waves of a photon are of course the ordinary light-waves of the undulatory theory. If the light is of frequency ν, the equation which governs their propagation is

$$\left(\frac{d^2}{dx^2} + \frac{d^2}{dy^2} + \frac{d^2}{dz^2}\right)\psi + \frac{4\pi^2\nu^2}{c^2}\psi = 0 \quad \ldots\ldots(\text{H}),$$

in which ψ is any component of electric or magnetic force or any linear combination of components. This can be shewn to be precisely identical *mutatis mutandis* with Schrödinger's equation (F) (p. 203) for the propagation of the electron waves of a single electron.

But when two photons meet, the equation of wave-propagation is not identical *mutatis mutandis* with Schrödinger's equation (G). In fact it is still equation (H). The reasons for the divergence here is of course that photons do not interact when they meet, whereas electrons do; this again can be traced back to the fact that photons do not carry electric charges as electrons do. Thus, although the waves of a million electrons need a space of three million dimensions for their proper representation, a space of three dimensions suffices for the waves of a million photons. Now we may properly identify this latter space with the space of our everyday life. For this is the space in which we see the sun, moon and stars, which again is the space in which the photons from the sun, moon and stars travel, and ultimately reach us. This is precisely the space of equation (H).

Thus we may say that photon waves can be represented

in our ordinary everyday space—this indeed is the definition of this space—but that electron waves cannot be so represented.

Let us, however, proceed somewhat further with our comparison or contrast of electron waves with photon waves. On p. 160 we discussed an experiment in which monochromatic light was passed through two pin-holes and made to form a pattern of light and dark bands by interference. If the pattern is formed on a sensitive photographic plate, chemical action will occur at places where the pattern is light, but none where it is dark. Let us now picture our monochromatic light as a shower of photons, all having the same wave-length and so also the same momentum. We have seen that photons do not admit of localisation at single points of space; they are merely free vibrations of the laboratory (p. 167) or else the waves of such vibrations combined into wave packets (p. 206).

Suppose, however, that we press the bullet aspect of radiation to the illegitimate extreme of supposing that each photon can be localised at a particular point of space. Then, to make our picture consistent, we must suppose that no photons fall on our screen where the interference pattern is absolutely dark; they all fall where it is light. Indeed, we must suppose that the number which falls on any small area of the screen is precisely proportional to the total illumination of the screen. We can even dispense with the screen altogether, and speak of the number of photons in a small volume of empty space; this number will of course be proportional to the total amount of light-energy in the volume of space in question. To see that this is a legitimate and necessary extension of our ideas, we need only imagine a small screen placed at the far end—the end away from the light—of the small piece of space in question. This will catch the photons, much as a fish-net catches the fish swimming into it, and it is easy to see that the light-energy per unit volume of space is exactly proportional to the number of photons in this unit volume.

Electrical theory, however, teaches us to regard energy as being spread continuously through space, not concentrated in the isolated points which happen to be occupied by photons. How are we to reconcile this with our supposition that the energy is merely the aggregate energy of individual photons occurring at isolated points?

The ordinary theory of gases (p. 157) points the way. It shews us a gas as a number of bullet-like projectiles—its molecules. The mass of the gas is concentrated wholly in the few points of space which happen to be occupied by molecules. Nevertheless when we speak of the density of the gas, we suddenly change our picture; we form, so to speak, an out-of-focus picture in our minds in which the separate molecules are blurred into a continuous cloud, and what we describe as the density of the gas is merely the density of this blurred cloud, which we see spread continuously through space.

When we put this picture back into sharp focus, we see the separate molecules again. We see that the true density of matter varies abruptly from point to point—it is large here, where there happens to be a molecule, and zero at an adjacent point where there is no molecule. Yet our former conception of a density which varied continuously from point to point still retains a perfectly precise and clear-cut meaning. It is this: if we take a tiny fragment of space surrounding a point P, the chance of our finding a molecule inside it is exactly proportional to the density at P.

So, when we picture a beam of light as a shower of bullet-like photons, we must suppose that the density of light-energy at each point of space gives a measure of the chance of our finding a photon there. In ordinary electrical theory the density of light-energy is shewn to be $E^2 + H^2$, where E and H are the electrical and magnetic forces measured in suitable units.* Thus in our photon-picture of light, we may interpret E and H as quantities which between them give a measure of the probability of our finding a photon at a particular spot in space.

If we take Schrödinger's equation which governs the propagation of electron waves, and make the changes appropriate to the transition from an electron to a photon, we obtain, as we have seen, the equation of propagation of electric disturbances, of which waves of light form a special case. Thus Schrödinger's quantity ψ must in some way be analogous to the quantities which specify electric disturbance—i.e. to electric and magnetic force. These latter provide us with a measure of the probability of finding a photon at a particular spot when we picture photons as being localised at points of space, and in the same way ψ must provide a measure of the probability of finding an electron at a particular spot when we picture electrons as being localised at points.

Mathematical theory discloses the exact relation between

* In the more ordinary units, the density of energy is of course $(E^2 + H^2)/8\pi$.

ψ and the probability in question. The propagation of electric disturbance is determined by the equation

$$\left(j_1\frac{d}{dx}+j_2\frac{d}{dy}+j_3\frac{d}{dz}+j_4\frac{d}{d\tau}\right)(E+iH)=0,$$

where E and H are the electric and magnetic force,* i is the square root of -1, and j_1, j_2, j_3, j_4 are symbols whose exact meaning does not matter at the moment. They are a sort of geometrical square roots of -1, being unit vectors drawn in the directions of the axes of x, y, z and τ respectively, where $\tau = ict$.

We may compare this with the equation for the propagation of electron waves, which Dirac has reduced to the form

$$\left(E_1\frac{d}{dx}+E_2\frac{d}{dy}+E_3\frac{d}{dz}+E_4\frac{d}{d\tau}-m\right)\psi=0.$$

Here ψ is the symbol which appears in Schrödinger's equation, and m is the mass of the electron, this introducing a new term which did not appear in the photon equation above. Again E_1, E_2, E_3, E_4 are symbols whose exact meaning does not matter at the moment. Like j_1, j_2, j_3, j_4 they are square roots of -1, but being matrices they no longer admit of simple geometrical interpretation. This re-affirms the fact already noticed (p. 216), that photon waves can be represented in space and time, whereas electron waves cannot.

A comparison of the two equations immediately suggests that ψ for an electron is the analogue of $E + iH$ for

* Actually $E + iH$ stands for the six-vector $X + i\alpha$, $Y + i\beta$, $Z + i\gamma$, $X + i\alpha$, $Y + i\beta$, $Z + i\gamma$, and the equation must be read as a vector equation. The vanishing of the four components of the vector on the left then give Maxwell's eight equations exactly.

a photon. We have already seen (p. 181) that ψ will have an imaginary part as well as a real part, so that we may replace ψ by $\psi_1 + i\psi_2$. Then ψ_1 is analogous to E and ψ_2 to H, so that $\psi_1{}^2 + \psi_2{}^2$ is analogous to $E^2 + H^2$ which, as we saw, gives a measure of the chance of finding a photon at a particular spot of space, when we picture photons as existing at spots in space.

Waves of Probability

Reasons of this kind led Born to conjecture that, if we try to locate electrons at spots in space, the value of $\psi_1{}^2 + \psi_2{}^2$ at any particular spot in space will give a measure of the probability of our finding an electron there. Such a statistical interpretation is in keeping with the statistical origin of the wave-mechanics. As Heisenberg's relation was obtained from statistical data, our working model of this relation is *a priori* likely to be only of statistical significance.

We can test the truth of the foregoing conjecture in various ways, and it emerges triumphantly from all.

To fix our ideas, let us return to the experiments we have already discussed, in which a current of electricity passes through a thin metal film, and emerges on the other side with the attributes of waves. We have to suppose, as on p. 201, that the oscillation-characteristic was inherent in the electrons of the current from the outset, and that as soon as the stream impinged on the film these oscillations gave rise to waves. The stream accordingly emerged from the film with the characteristics of a group of waves. There would be crests and troughs giving rise to places of great wave-intensity, and quiescent regions of small or zero wave-intensity. If we picture the current as a shower of point-electrons, then there would be many electrons at places where the wave-intensity was great, few electrons where it was small, and no electrons at all at places of complete quiescence, and the same would remain true after the current had emerged from the film. This exactly explains the pattern formed on the photographic plate.

Now let us gradually reduce the strength of our electric current until it almost vanishes, and let us consider what happens in an interval of time so short that only one electron passes. Broken electrons and fractions of electrons are never found in nature, so that to keep our picture consistent with the known facts of nature, we must suppose that our single electron is not broken up by the experiment, but retains its identity, and emerges from the experiment, as it went in, a single particle charged with electricity. Thus it can only interact with a single particle of the sensitised plate at one single point. It cannot form a complete pattern; only a shower of electrons can do this. The light and dark of the pattern previously formed by a whole shower have, however, given a sort of graphical representation of the probabilities of any individual electron striking particular spots. Thus when our solitary electron comes along, there is no chance at all of its striking a spot where the previous pattern was dark; a million million electrons have already taken their chances of hitting such a spot, and not a single one succeeded in doing so, so that the chance that this isolated traveller will do so is *nil*. But there is a finite chance that the electron will strike any area which was bright in the previous pattern, and this chance is proportional to the total brightness of the area in question. Before performing the experiment with the single electron, we can only say that the chances of such and such a result are so and so, the various probabilities being determined by the waves specified by the Schrödinger equation.

In this way electron waves become reduced to mere diagrammatic representations of probabilities, and this explains at once why they need a space having three times as many dimensions as there are electrons. When, however, there is only one electron, or a shower of electrons whose motion is indistinguishable from one another, only three dimensions are needed for our diagram, and it is natural to think of this as constructed in ordinary space. In this sense we may think of the Schrödinger waves of a single electron as existing in ordinary space, although we must always remember that they are mere mathematical waves, and possess no physical reality.

Ordinary material waves, such as waves of sound or ripples on the surface of water, spread their energy about until finally it is diffused through the whole of the space accessible to the waves. The total amount of energy remains the same throughout, the process of wave-propagation merely altering its distribution in space.

The conjecture we are now considering endows electron waves with an analogous property. On passing from material waves to electron waves, we replace energy by the chance of finding an electron. Just as the total energy of the material wave remained constant throughout, so of course does the total probability in the electron wave, since the aggregate of all the probabilities at the different points of space must always be exactly equal to the total number of electrons.

When a material wave meets a surface of another substance, part of it may be reflected and part of it transmitted, the total energy of the two new waves being exactly equal to that of the old. In the same way, when an electron meets a material surface, its probability wave breaks up into two parts, a transmitted wave and a reflected wave, which represent the probabilities of the electron being transmitted and reflected respectively. Just as the total energy remains unaffected in a material wave, so does the total probability here. This circumstance endows the Schrödinger waves with many of the attributes of material waves, although it naturally does not provide the slightest justification for supposing that they are material waves.

It may at first seem strange that a probability should be propagated in so distinct a wave-like form, with a definite wave-length and period of oscillation. It seems less strange when we notice that a probability field must conform to the theory of relativity and so may be treated precisely in the way in which we treated the supposed structure of an electron on p. 204. With this in mind, we see at once that probabilities must necessarily be propagated in waves, the whole group of waves travelling with exactly the known speed of the electron.

We do not of course solve the whole problem of the nature of these waves by describing them as waves of probability; before our task is finished, we must specify the meaning of probability with far greater precision than has yet been done.

The mention of a probability in ordinary life implies

that our knowledge is in some way imperfect. We speak, let us say, of the "probability" of a good channel crossing as we travel in the train to Dover, although we should not do so if we knew the state of the sea. One man may say, "The sea is only rough one day in three this month, so that the odds are two to one that we shall have a good crossing". Another may say, "The odds are better than that, for the weather forecast predicts a smooth sea, and it is right 95 times out of 100". A third may say, "It is practically certain, for I saw this morning's telegram at the Meteorological Office saying there was a smooth sea and no wind". These estimates of probability are all different, and yet they can all be correct; this is possible because probability involves two things, a future event and present knowledge. We have interpreted the system of Schrödinger waves as giving a definite estimate of the probability of a future event, and must proceed to inquire: Relative to what present knowledge is this a true estimate of probability?

Let us first notice that as our knowledge increases, the probability of a smooth sea or any other event continually approximates to either zero or unity; it gradually changes into a certainty, either one way or another. The meteorological expert who is armed with the latest telegrams need hardly use the word probability at all; he can say with practical certainty either that the sea will be smooth, or that it will be rough.

In physics we may need to speak of the probability of a future event, or of the result of an experiment, for either of two reasons—it may be that we have inadequate knowledge of present conditions, or it may be that even when present conditions are fully known, there is still uncertainty about the future—in other words, the principle of the uniformity of nature may fail.

Subjective Probability

On the former alternative, the probability of a specified future happening, or result to an experiment, is a subjective probability; if we have different amounts of knowledge as to present conditions, your estimate of the probability may be different from mine, and yet we may both be right. On the latter alternative, the probability is wholly objective; even nature herself does not know the result of the experiment until after it has happened. The question "What is the probability of a specified result?" admits of only one answer.

The former was the only alternative which the nineteenth-century physicist would have admitted at all. He would have dismissed the second as preposterous, as indeed the scientific layman still does. Unaccustomed to thinking beyond the apparent determinism of the natural events which make up his everyday experience, he unconsciously assumes that a similar determinism must permeate nature even down to its most small-scale operations, and denounces any alternative as illogical or contrary to the laws of nature—often with much heat and emotion.

Let us again concentrate on the particular instance of the single electron shot at the thin metal film. The layman in science, like the physicist of the old school, would probably say its path would be determined by the obstacles it met in the film. If he played billiards, he would know that when one ball hits another a slight difference in the path before impact may make a very great difference in the path after impact. So he might argue that, as we could not know the exact circumstances of the collision between the electron and the atom or atoms which deflected it, we could not know the final path of the electron and so could only speak of the "probabilities" of one path or another.

This interpretation of probability will not stand scrutiny, since the wave-length underlying the pattern does not depend on the spacing of atoms in the film, but solely on the speed of the electron.

Objective Probability

The second alternative, which postulates indeterminacy in nature, although foreign to nineteenth-century thought, has a far longer history than the former. When the more intelligent of our remote ancestors said that the prospects of a fair crossing depended on the whims of Poseidon and Boreas, they did little more than personify nature and attribute indeterminacy to her. Even up to the time of Newton, the concept of indeterminacy played a large part in science. The experiment of letting a shower of electrons fall on a metal film has its optical counterpart in the falling of a beam of light on the surface of a transparent substance. Some of the light is reflected, and some transmitted, so that when moonlight falls on the surface of the sea, some enters our eyes and we see a reflection of the moon, while the remainder lights up the depths of the sea so that the fishes too can see the moon. If we picture the moonlight as a shower of photons, then clearly some photons must undergo reflection at the surface of the water, while others do not. If, however, the beam is reduced to a single photon, then since photons are indivisible the whole beam must go either one way or the other, and we shall only be able to speak of the "probability" of its being reflected or transmitted.

Newton, who regarded a beam of light as a shower of bullet-like corpuscles, encountered a similar difficulty, and met it by imagining that the molecules which formed the surface of the water suffered from "alternate fits of easy transmission and of easy reflection".

It is not clear what precise degree of absence of determinism may have been implied in this. Whatever it was, it left science when the corpuscular theory of light gave place to the undulatory, only to reappear in the present picture which envisages a beam of light as a shower of indivisible protons. Like its predecessor, the light-corpuscle, a photon may follow one path or another, but cannot distribute itself over two paths, and once again its choice becomes, to all appearances, a matter of probability.

Instances of a similar apparent want of determinism have recently appeared in other departments of physics. A conspicuous instance is provided by radio-active transformation. In 1903 Rutherford and Soddy found that radio-active substances disintegrate in a way they described as "spontaneous"—the rate of decay cannot be expedited or retarded by any known physical process. Each year a certain fraction of all the atoms of radium in the world disintegrate into simpler atoms, the individual atoms being to all appearances selected by pure chance and nothing else. If anything else could select them, it ought to be possible to concentrate the selecting agency on one special sample of radium and expedite its disintegration. So far no such selecting agency has been discovered, and theoretical considerations make it highly probable that, apart from extreme heat such as cannot be produced on earth, none such can exist.

In 1917 a theoretical investigation by Einstein seemed to shew that spontaneous processes of this kind must pervade the whole of nature. He began by supposing that atoms could only exist in certain distinct states—the suppositions which had previously been made by Bohr (p. 53), and subsequently received experimental confirmation in the experiments of Franck and Hertz—and that they absorbed or emitted energy by complete photons in passing from one state to another. He then shewed that ordinary temperature radiation (p. 151) could be interpreted as an assemblage of photons produced in this way, but only on certain conditions. Many of the photons observed in the radiation could be accounted for by the interaction between the radiation itself and the atoms of the substance, but Einstein shewed that a residue remained, which could only be accounted for on the supposition that the atoms fell spontaneously from one of their possible states to another. Thus even the familiar everyday phenomena of temperature radiation seem to call for some sort of action which is inconsistent with a strict determinism.

Einstein is of the opinion that these particular phenomena are consistent neither with indeterminism nor with causality as at present understood. He says:*

"Indeterminism is quite an illogical concept.... If I say

* *Where is Science going?* by Max Planck (1933), pp. 202, 203.

that the average life-span of a radioactive atom is such and such, that is a statement which expresses a certain order (*Gesetzlichkeit*). But this idea does not of itself involve the idea of causation. We call it the law of averages; but not every such law need have a causal significance. At the same time if I say that the average life-span of such an atom is indetermined in the sense of being not caused, then I am talking nonsense....

"When Aristotle and the scholastics defined what they meant by a cause, the idea of objective experiment in the scientific sense had not yet arisen. Therefore they were content with defining the metaphysical concept of cause. And the same is true of Kant. Newton himself seems to have realized that this pre-scientific formulation of the causal principle would prove insufficient for modern physics....Now I believe that events in nature are controlled by a much stricter and more closely binding law than we suspect to-day, when we speak of one event being the *cause* of another. Our concept here is confined to one happening within one time-section. It is dissected from the whole process. Our present rough way of applying the causal principle is quite superficial. We are like a child who judges a poem by the rhyme and knows nothing of the rhythmic pattern. Or we are like a juvenile learner at the piano, just relating one note to that which immediately precedes or follows. To an extent this may be very well when one is dealing with very simple and primitive compositions; but it will not do for the interpretation of a Bach Fugue. Quantum physics has presented us with very complex processes and to meet them we must further enlarge and refine our concept of causality."

Professor Weyl of Göttingen, writing on the metaphysical implications of science, expresses the same opinion* :

"These considerations force upon us the impression that

* *The Open World* (1932), p. 43.

the law of causality as a principle of natural science is one incapable of formulation in a few words, and is not a self-contained exact law. Its content can in fact only be made clear in connection with a complete phenomenological description of how reality constitutes itself from the immediate data of consciousness."

Even though a certain measure of indeterminism may appear necessary to explain certain small-scale phenomena, the principle of the uniformity of nature still prevails so long as nature is only studied in appreciable amounts. Even the tiniest bit of matter we can perceive through our senses contains billions of atoms, and if each of these is free to go to the right or left as it pleases, the laws of probability secure that, as far as our senses can tell, half will go each way. For this reason, our everyday experience will never shew us any violations of the so-called law of the uniformity of nature, and the man whose thoughts are guided only by intuition or instinct, or who holds to the common-sense view of nature, is certain to be a determinist.

We shall return to a discussion of this question after obtaining further evidence in the next chapter.

CHAPTER VII

INDETERMINACY

We have seen that our whole knowledge of the external world of physics may be pictured as arising from the impact of photons of energy either on our sense organs or on our physical instruments. As these photons occur in such profusion and variety, it might have been hoped that they would give us an almost perfect knowledge of the outer world.

Yet, as a means of acquiring knowledge, photons suffer from one very serious limitation. They are indivisible; no experiment has ever revealed a fraction of a photon or given any reason for supposing that energy can be either emitted or absorbed in fractions of photons. Thus the only means which are at our disposal for the study of physical nature suffer from a certain coarse-grainedness.

This is of little consequence as regards direct study by our senses, since these are even more coarse-grained. Each sense has its perceptions limited by a certain "threshold of sensation", and if the stimulus of a physical effect falls below this, the sense in question registers nothing at all. We cannot experience the sweetness of a single molecule of sugar, nor the smell of a single molecule of musk; neither can we hear a bell at more than a certain limit of distance, nor see a star which is below a certain limit of faintness. In general we cannot experience a single photon; thousands at least are necessary to attain the threshold of sensation.

Our physical instruments have in a sense a similar "threshold of sensation", this being the arrival of a single complete photon. Like all other physical structures, they accept energy and momentum only by complete photons.

The Uncertainty Principle

Thus the most refined piece of information we can obtain about any piece of the universe is that conveyed by the arrival of a single photon. This transfers to our instrument energy and momentum which it has brought with it from the fragment of the external world in which it originated. Now just as a shot gives a backward kick to the gun from which it is fired, so a photon gives a backward kick to the atom which sends it out, and through this atom to the fragment of the universe which we are trying to study. Thus it may give us accurate news of the universe as it was, but the kick it gave to the universe in leaving it to bring us news makes the news out of date before it reaches us; we receive news only of an old universe which has already passed away.

It might be thought that as photons carry all possible amounts of momentum ranging from zero upwards, we could obtain as accurate information as we pleased by employing photons of small momentum. This is, in the abstract, true; in practise it only shifts the difficulty. For photons of small momentum have such long periods of oscillation that we cannot fix the instant to which their information refers with any great precision; it is like trying to time a hundred yards race with a grandfather clock that only ticks seconds.

Thus we are confronted with the dilemma that one

kind of photons are so energetic that they give the universe a violent kick before leaving it, and so give us inexact information about its present condition, while the other kind are so slow in telling their story when they arrive that they cannot give us exact information in respect of time. Intermediate kinds fail in both respects.

Science has found no way out of this dilemma. On the contrary, it has proved that there is no way out. What is known as Heisenberg's "principle of indeterminacy" or "uncertainty principle" shews that so long as we can only explore nature by complete photons, there is no hope of obtaining information which is perfectly exact with respect to both time and space. Exactness in either direction is obtained at the price of inexactness in the other; we can only prevent the shoe pinching at one place by letting it pinch at another.

Exact mathematical discussion shews that, as we try one kind of photon after another, the product of the two errors can never fall below a certain minimum value. If the experiments are designed and performed with perfect skill, the product of the two errors is the same for all kinds of photons, and is equal to this minimum value.

For instance, to obtain a complete knowledge of the motion of a particle, we need two data—the exact instant at which the particle passes an assigned landmark in our apparatus, and the exact speed with which it is moving as it passes this landmark. If we agree to measure the speed of our particle in terms of its momentum, then we find that the product of the errors in position and momentum can never be less than Planck's constant h. We have already noticed how this quantity dominates the

whole of atomic physics. We come upon it here as specifying the coarseness of the probe, the photon, with which
we are trying to penetrate the outer world.

In centimetre-gramme-second units, the value of h is
$6·55 \times 10^{-27}$ and the mass of an electron is 9×10^{-27}.
Thus the product of the uncertainties in the position and
speed of an electron, measured in the same units, is $0·73$.
For instance, if, by letting it make a flash on a screen, or
by any other means, I discover that an electron is within
a hundredth of a centimetre of a certain point, then the
speed of its motion will necessarily be uncertain to at
least 73 centimetres a second—the rate of a slow walk.

We have so far pictured an electron as a particle, but
we can also picture it in terms of Schrödinger waves. If
the two pictures specify the same object, we ought of
course to be able to derive precisely the same "principle
of uncertainty" from the wave picture as we have already
derived from the particle picture. We shall now see that
this can be done.

When we regard an electron as a system of waves, their
wave-length depends on the speed of the electron in the
way already described. Thus the problem of measuring
this speed with exactness reduces to that of specifying a
wave-length with exactness. Abstract mathematics shews
that this cannot be done unless we have an infinite number
of waves at our disposal; with fewer waves the concept
of wave-length has no exact meaning. If we have a million
waves at our disposal, we can measure their wave-length
to within something like a millionth part of its amount,
but to speak of measuring it more accurately than this is
meaningless.

We can illustrate this point of pure mathematics by the

difficulties which arise when we try to measure the wave-length of a finite train of waves in the laboratory. For simplicity let us suppose they are wireless waves, and that we allow them to fall on an ordinary wireless receiving set, which can be tuned to any wave-length we please. Any train of waves will set up disturbance by resonance over a wave-band of finite width. As we lengthen the train of waves, the interference with neighbouring wave-lengths diminishes, but it does not completely disappear until the train of waves is made infinitely long. Only then can we say that the waves have a clearly defined wave-length.

It follows that we cannot specify the speed of motion of an electron with perfect precision unless it is represented by an infinitely long train of waves. But since we have seen that the waves represent the probabilities of finding the electron in different positions in space, the electron may be anywhere along the whole length of the train, and an infinitely long train of waves implies an uncertainty of infinite amount as to the position of the electron.

Let us now pass to the other extreme, and imagine an infinitely short train of waves to pass over the receiving set. The set sees nothing in such a train of waves but a mere sudden disturbance, which disappears the instant it has come into being. As every wireless expert knows, such a disturbance affects all wave-lengths indiscriminately, and so cannot be said itself to have any definite wave-length. An infinitely short train of waves of this kind represents an electron whose position can be specified with precision, but we see that its wave-length, and so also the momentum and speed of its motion, are completely indefinite.

A mathematical discussion of intermediate cases leads

to exactly the principle of indeterminacy already explained. Greater precision in momentum implies greater uncertainty in position, and vice-versa; the product of the two uncertainties can never be less than Planck's constant h, and, under the most favourable circumstances, is exactly equal to h.

Interpretation of the Wave Picture

It is not surprising that the particle picture and the wave picture lead to the same "principle of uncertainty"; there would have been something wrong had they not done so. Yet they lead to this principle in very different ways. When we use the particle picture of an electron, the uncertainty refers to the knowledge of nature we obtain through experiments on nature. When we use the wave picture, we find that the uncertainty is inherent in the picture itself. In brief, the particle picture tells us that our knowledge of an electron is indeterminate; the wave picture that the electron itself is indeterminate, regardless of whether experiments are performed upon it or not.

Yet the content of the uncertainty principle must be exactly the same in the two cases. There is only one way of making it so; we must suppose that the wave picture provides a representation, not of objective nature, but only of our knowledge of nature.

Earlier in our book we saw nineteenth-century science trying to explore nature as the explorer explores the desert from an aeroplane. The uncertainty principle makes it clear that nature cannot be explored in this detached way; we can only explore it by tramping over it and disturbing it; and our vision of nature includes the clouds of dust we ourselves kick up. We may make clouds of different kinds, but the uncertainty principle shews that there is no way of crossing the desert without raising a cloud of some kind or other to obstruct our view. The wave picture depicts the blurred view of nature that we see through these dust clouds, so that, as we shall shortly see, there are as many wave pictures as there are ways of raising a dust cloud.

If we turn our thoughts back to the origins of the wave picture, we can see why all this must be. This picture was introduced to provide us with a sort of working model of Heisenberg's equation, and this equation was concerned solely with observables—that is to say, not with objective nature but with our observation of nature. Heisenberg attacked the enigma of the physical universe by giving up the main enigma—the nature of the objective universe—as insoluble, and concentrating on the minor puzzle of co-ordinating our observations of the universe. Thus it is not surprising that the wave picture which finally emerged should prove to be concerned solely with our knowledge of the universe as obtained through our observations.

Electron Waves as Waves of Probability

This interpretation of the wave picture explains a great deal that would otherwise seem very mysterious, and gives greater precision to the discussions of our previous chapter.

We there described the waves of the wave picture as "waves of probability", but were unable to assign any precise meaning to the term. Let us now suppose that we perform an experiment to find the speed and position of a moving electron. An experiment of one type may fix its position with great accuracy but its speed with great uncertainty; the electron so observed appears in the wave picture as a short train of waves. An experiment of another type may fix the speed with great accuracy but the position with great inaccuracy; the electron is now represented by a long train of waves. The same electron may be represented by two different wave pictures, not because it is itself different in the two cases, but because our knowledge of it is different in the two cases. Thus the waves represent subjective probabilities.

Suppose our experiments tell us that an electron is at such-or-such a point in space, subject to a certain indeterminacy, and is moving with such-or-such a speed, again subject to a certain indeterminacy.

We may represent our lack of precise knowledge as to the position of the electron by substituting a fog for the latter; our knowledge is that the electron is somewhere inside the fog. If we knew the precise speed of the electron, we could let the fog move forward at this precise speed, and the electron would always lie inside the fog. We cannot, however, know the precise speed, but only know that it lies within certain limits, as for instance between 50 and 55 miles a second. To represent this, we must regard our fog as being made up by the superposition of a number of separate individual fogs, and let these individual fogs move forward at the various speeds within these limits—one at 50 miles a second, another at 51 miles a second and so on. We shall then know that at every instant the electron lies somewhere within the fogginess produced by all these fogs. Let us notice that the area of fogginess continually increases; this means that our knowledge of the position of the electron continually gets more and more vague.

This is faithfully represented in the wave picture, because it is a general property of waves to spread. The electron itself retains its identity throughout, but the fog—representing not the electron, but our knowledge of it—must perforce continue to spread until it ultimately pervades all space.

For many purposes the weight of a massive body may be supposed concentrated in a single point, which we call the centre of gravity of the body. So also, for many purposes, we may suppose the whole electron to be concentrated at the centre of gravity of the fog. For instance, Ehrenfest has shewn that, when the waves of the fog travel as directed by Schrödinger's equation, the centre of gravity of the fog will describe precisely the same curved path in an electric field as a single point-electron would describe.

Let us now imagine that our moving electron meets a metal film. Each small fragment of our fog will break up into a system of waves, in the way we have already explained, and the aggregate of all these systems of waves constitutes the wave pattern of the electron, such as we see in fig. 2 of the frontispiece. This new wave pattern is more extended in space than the original fog, because our uncertainty as to the position and motion of the electron has been increased by a further uncertainty as to its conduct in passing through the metal film.

Light-waves as Waves of Probability

We can discuss light and light-waves in a precisely similar manner. We can picture light as consisting of photons, and these photons have wave pictures just as electrons have; they are neither more nor less than the ordinary waves of the undulatory theory of light. When we picture photons as localised at points, we have already seen (p. 218) that these waves must be interpreted as waves of probability—the probability of finding the photon at a given spot. It is rather surprising to discover that we must take the further step of regarding them as mere diagrammatic representations of our knowledge as to the whereabouts of photons, yet a simple instance, which has been discussed by Einstein and Ehrenfest, will shew that such is the case.

A photon which meets the surface of a transparent substance may be either reflected or transmitted. For the sake of simplicity, let us suppose that the chances of the two events are equal. This means that when the wave-system of the photon falls on the reflecting surface, it will divide into two beams of equal intensity, one reflected and one transmitted. These are of course beams of ordinary everyday light. After a few seconds interval, these two beams may be a million miles apart, which means that on our present knowledge we cannot fix the position of our photon to within a million miles.

A new experiment will, however, clear up some at least of our uncertainty. Let us calculate the path of the reflected ray by geometrical means, and place across it a screen which will register a spot of light if the photon strikes it. Up to this moment, our knowledge of the doings of the photon has been represented by two beams of light-waves; one is just about to fall on the screen, the other is a million miles away. We watch the screen. If it does not light up, we know that the photon has chosen the other path, and is a million miles away. If it does light up, we see that the photon has chosen this particular path, and this new knowledge completely transforms the system of waves. We now know for certain that the photon is not at any point on the distant beam, because it is here, and the whole system of distant waves disappears for ever; it is completely annihilated. On the other hand, the beam of light we are watching becomes contracted to a point— the point which lighted up on our screen.

It is at first very startling and not a little puzzling to reflect that by the mere act of watching a screen here we can annihilate light-waves a million miles away. Old-fashioned physics told us that light-waves were waves of energy, so that the act of looking at a screen here has apparently destroyed energy a million miles away. Even if the total energy is conserved, our action has removed the energy from there to here, and this at infinite speed, although we used to be told that energy could not travel faster than light.

The paradox disappears as soon as we treat the light-waves as waves of probability, their extension in space defining the uncertainties of our knowledge. The waves are no longer waves of energy, but of the chance of finding energy. When there are billions of billions of photons in the field, the total measure of the chance of finding energy is, for all practical purposes, the same thing as the measure of the energy we shall find, and we need not trouble to distinguish between the two. When there is only one photon involved, the distinction becomes important. What is transferred now is not energy, but the chance of finding energy, which in turn depends on our knowledge as to the whereabouts of the energy. This may well be transferred, not with the speed of light which is finite, but with the speed of thought, which is infinite.

As this is one of the most difficult parts of the new quantum theory, let us try to illustrate it by a very prosaic illustration. Suppose I am anxious to meet my relative John Smith, who is owing me a sum of money, and that all I know of him for certain is that he left his home in London three days ago for an unspecified destination. My knowledge as to the whereabouts of John Smith is represented by a fog which extends over all those parts of the earth's surface which are within three days' travel of London. I next find that a passenger named John Smith sailed on the *Majestic* three days ago for New York, and the fog becomes particularly dense in mid-Atlantic, three days out from land. I hurry to a cable office to communicate with the *Majestic* in the hope of getting a reply which will inform me, with the speed of light, whether my relative is in mid-Atlantic or not. But, on my way, I run into John Smith himself. This simple act not only concentrates all the fog into one spot in space, namely that at which my relative is standing; it also abolishes the fog in the Atlantic, and does this far more promptly than a wireless message, travelling with the speed of light, could do. It can do this because the fog is not a material fog, such as delays shipping; it consists of knowledge—knowledge about John Smith.

So, in the last resort, the waves which we describe as light-waves, and those other waves which we interpret as the waves of an electron and a proton, also consist of knowledge—knowledge about photons, electrons and protons respectively. We can see now why modern science does not need the old material ether, millions of times more dense than lead, for light-waves to travel through.

The Waves of the Hydrogen Atom

Let us now revert to electron waves, and consider what happens when an electron, originally moving freely through space, combines with a proton to form a hydrogen atom. If we knew its original path with fair accuracy, the electron may be represented by a fairly compact packet of waves when it first comes under the attraction of the proton. In accordance with the principles already explained, this wave packet will describe a curved path round the proton (p. 239), and will continually increase in extent as it does so (p. 239). The general principle that waves continually spread shews that there is only one end possible—the wave packet must ultimately fill the whole of space. Throughout all these changes, as also when the final state is reached, the electron waves will conform to Schrödinger's equation. This equation has many solutions, which will of course represent many different kinds of waves. Some will represent permanent unchanging systems of waves, and these are found to specify the possible permanent states of the hydrogen atom.

It is a comparatively simple problem to discover all the solutions of this type. Such solutions are found to exist for certain values of the energy E in equation (F) of p. 203 and for no others. Thus the hydrogen atom can only exist permanently in certain discrete states—specified by their different amounts of energy—precisely as was first postulated by Bohr, and was subsequently confirmed by the experiments of Franck and Hertz. These amounts of energy are of course easily calculated, and the lines of the hydrogen spectrum are found to correspond exactly to transitions from one of them to another, so that the triumph of wave-mechanics is complete.

It may be of interest to try to understand something of the geometrical disposition of these waves. Bohr's earlier theory, as we have already seen (p. 53), supposed the hydrogen atom to consist of a charged particle—an electron—describing a circular orbit round another charged particle—the nucleus. The electron was supposed to be confined to orbits possessing certain definite amounts of energy, and so also definite diameters; it was as though certain grooves were cut in space, and the electron was compelled to run round and round in the same groove except on the comparatively rare occasions when it jumped from one groove to another.

Wave-mechanics also finds that the electron-nucleus combination can only have these particular amounts of energy. But the electron is no longer a particle; it is a system of waves running round and round the nucleus, somewhat as waves might run round and round in a circular trough of water, except for the quite important difference that troughs of water have clearly defined boundaries, whereas these waves have not. If, notwithstanding this, we like to imagine the waves imprisoned in troughs, then we find that these troughs must be of certain definite diameters such that the complete circumference of a trough may be occupied by one, two, three or any other exact number of complete waves, but never by a fractional number of waves. This condition makes the diameters of the troughs very nearly the same as the diameters of the orbits which Bohr had previously calculated from his simpler theory, namely, 1, 4, 9, 16, ... times the diameter of the normal atom in its normal state of lowest energy.

With a view to understanding this, let us consider a very simple hypothetical hydrogen atom which admittedly has not much relation to actual facts. Let us suppose that its electron is in some way constrained—rather as imagined in Bohr's first theory—perpetually to move round the nucleus in a circle of always the same radius a. If the atom is beyond the reach of our experiments, as for instance an atom in Sirius, our knowledge of the position of its electron will consist of the single fact that this is at a distance a from the nucleus. We cannot know which point of its orbit it will occupy at any instant, and neither can we know the orientation of this orbit in space. Thus our knowledge of the position of the electron is represented by a thin shell of fog, forming a sphere of radius a surrounding the nucleus.

If we suppose that the electron is constrained to move in one or other of a number of circular orbits, having radii a, b, c, ..., our knowledge will be represented by a number of thin shells of fog having radii a, b, c, ...

Let us now go one step farther in the direction of reality. An electron which is describing a circular orbit of radius a may be deflected, without loss or gain of energy, into an elliptical orbit, in describing which it will be alternately inside and outside its original circular orbit. Its distance from the nucleus will range between $a(1+e)$ and $a(1-e)$, where e is the quantity known as the eccentricity of the ellipse. As this can have any value between 0 and 1, the distance of the electron from the nucleus may be anything between 0 and $2a$. Thus if we know nothing about the electron except that it is describing an orbit of specified energy, our knowledge will be represented by a fog, which will extend through the whole of a sphere of radius $2a$, but no farther, and will be particularly dense at a distance a from the nucleus. If the electron can describe an orbit having any one of a number of specified energies, our knowledge will be represented by the superposition of a number of fogs lying inside the spheres of corresponding radii.

Such a system of fogs does not differ very widely from the probability diagram furnished by the wave-mechanics for the actual hydrogen atom, except for one very important point of difference. An electron moving in an orbit of specified energy can never move to more than a certain distance from the nucleus, whereas the probability diagram of the wave-mechanics extends throughout the whole of space. In brief, this diagram tells us that there is always a finite probability that the electron will reach points which would be entirely beyond its reach, because of insufficient energy, if it were an ordinary charged particle moving in space and time. This illustrates a complication which is not peculiar to the hydrogen spectrum, nor even to the more general problem of atomic structure, but permeates the whole of the new quantum theory. It seems as though, if an electron waits long enough, it will always be able to violate the law of conservation of energy by reaching places which its energy does not entitle it to reach. Gamow has suggested that the disintegration of radio-active nuclei may be due to this cause.

This makes it clear that if the conservation of energy is to remain in our picture of nature, we must attach a wider meaning to energy than we have done hitherto. We have so far thought of the energy of an electron as due to the position of a particle in space and its motion through space, and this in spite of our having seen that an electron, when inside an atom, cannot be represented in space and time. There is no obvious difficulty in taking a wider view of energy, and imagining that the electron can draw on energy which also cannot be represented in time and space. Heisenberg's equation has already shewn that the real electron has a greater complexity than mere position in space; if we know that its total energy has a certain value, and try from this to find its position in space, the complexity of its wave pattern shews us how many answers we may receive to our question. We may regard this wave pattern as representing the projection into space and time of all the configurations which have a given total energy.

Objective and Subjective Waves

Yet clearly this wave pattern is wholly objective, and at first this may seem to be at variance with our earlier statements that a wave pattern represented subjective knowledge. But it is easy to see that there is no conflict, and that our former interpretation of the electron waves as diagrammatic representations of subjective knowledge leads to precisely the result we have just reached. For, although we may at first have fairly precise knowledge which limits the position of an electron to a small region of space, yet uncertainties increase with the passage of time, so that our knowledge of this position gets continually vaguer; the electron waves spread. Finally, after a time which we may treat as infinite, they fill all space. And as the position from which an electron started an infinity of time ago can have no possible influence on the positions it may occupy now, the present electron waves are entirely independent of our knowledge, and so form an objective system.

We can illustrate this very simply by considering an electron in an otherwise empty universe. The wave-equation for such an electron is formally similar to the equation which governs the flow of heat in a solid of infinite extent. Thus our uncertainty of knowledge, which was at first localised within a small volume of space, spreads like heat through a solid. Just as the solid finally reaches a state of uniform temperature, which is independent of the spot from which the heat started, so the system of waves in wave-mechanics finally attains uniform intensity everywhere throughout space, and this independently of the earlier movements of the electron. This merely means that, no matter where the electron started, all positions for it are equally likely after an infinite time. In fact we could only predict its position after infinite time if we had known its original position and speed with infinite precision, and the uncertainty principle keeps this pair of data for ever beyond our reach. This final system of waves is of course wholly objective; it cannot represent subjective knowledge for the simple reason that we no longer have any. Or, to put the same thing in another way, it represents the fact that our knowledge is *nil*.

Another very simple solution of the wave-equation is of the form

$$e^{\frac{2\pi i}{\lambda}(lx + my + nz - Vt)}$$

and this represents waves advancing in the direction l, m, n, with speed V. This may be taken to represent an electron, or a current of electricity, advancing in the direction l, m, n, with a speed c^2/V. In choosing this solution to represent any particular electron, we assume we know that the electron is travelling with components of velocity which are precisely lc^2/V, etc. The uncertainty principle now tells us that we must pay for this precision in our knowledge of the momentum of the electron by accepting an infinite uncertainty in the position of the electron, which of course is precisely what the solution also tells us, since the waves are of uniform intensity throughout space. Here again we have a wholly objective system of waves.

Such systems of waves, extending uniformly through the whole of space, provide the only strictly objective representations of electrons or electric currents. Combinations of them give the concentrated wave packets which we observe and designate as electrons—the electrons of our observation. But if we ask mathematics to tell us how to build up such a wave packet—i.e. to provide us with the wave specification of a single electron moving freely in space—there is no answer. Or rather, strictly speaking, the answer takes the form of a further question—"Tell me first how much you know about the electron, and I will answer your question. If you know nothing, my only answer will be that I too know nothing". All the observational knowledge which has been put into the quantum theory and wave-mechanics proves to be of no avail to elicit what an electron is objectively.

The situation changes as soon as we put a second body,

as for instance a proton, into this space. Since the two particles attract one another, the electron, quite apart from any detailed knowledge on our part, is more likely to be in the neighbourhood of the proton than elsewhere. The wave-system now wraps itself symmetrically round the proton, displaying the various possibilities and relative likelihoods that we have already discussed. We now have an objective system of waves, and it appears that an electron can only be objectively specified when it is anchored to a proton or other material frame of reference; otherwise it merely fills all space uniformly. An objective electron localised in an empty universe is as meaningless as objective time—or rather we can attach to either as many meanings as we please.

Thus, if our wave picture of nature is to be wholly objective, it must contain no reference to isolated electrons or protons, but only to such combinations of these as can produce events which can affect our senses. On limiting our picture to these, we obtain a system of waves which is completely objective, in the sense that we must imagine them existing whether we experiment to discover their existence or not. The waves do not admit of representation in space and time, and so cannot be said to possess any physical reality.

Yet, in spite of this want of physical reality, this wave picture is in many respects more true to nature, and so is presumably more fundamental, than the particle picture which depicts nature as concrete objects existing in space and time. This is especially true of the more refined problems of atomic structure and spectral lines. Just as, in optics, the ray picture gives a rough approximation, but a wave picture is needed to exhibit the finer details of phenomena, so here the particle picture will often give a rough approximation to a truth which the wave picture explains with perfect precision. The following illustrations may serve to typify a whole mass of highly technical knowledge.

The spectra of the alkali metals consist entirely of "doubled" lines—pairs of lines which are quite distinct and yet very close together—such as the well-known D lines of sodium. Two Dutch physicists, Uhlenbeck and Goudsmit, tried to explain this doubling by supposing that the electrons of the particle picture spun round on their axes as they described their orbits in the atom. This gave the electrons slightly different energies according as they spun in the same direction as their orbital rotation, like the earth, moon and planets, or in the opposite direction, like the outermost satellites of Jupiter and Saturn, and the single satellite of Neptune. Further, the amounts of spin required to account for the doubling of the spectral lines agreed exactly with that needed to account for the Zeeman effect—the rearrangement of spectral lines that occurs when the incandescent gas is placed between the poles of a powerful magnet.

Thus the particle picture could be made true to nature by supposing the electrons to be spinning. Nevertheless such spins seemed very artificial until it was found, a few weeks later, that they were a necessary consequence of the wave picture. Just as light-waves admit of different kinds of polarisation, which we can represent in the particle picture of light by different spins of the photons (p. 159), so electron waves admit of different kinds of polarisation, which we may represent in the particle picture of matter by different spins of the electrons.

Yet this does not place the two pictures quite on the same footing. For we see that if the wave picture of matter is fundamental, all the electrons in the particle picture must necessarily be spinning. On the other hand, if the particle picture is fundamental—i.e. if nature really consists of particles, and the waves merely provide a diagrammatic representation of our imperfect knowledge of their positions—there is no obvious reason why the particles should spin at all. Both pictures explain the facts, but the explanation of the particle picture appears artificial, while that of the wave picture appears natural, and indeed inevitable.

At a later stage the experiments of Stern and Gerlach provided what amounts to an *experimentum crucis*, enabling us to decide between the particle picture and the wave picture. In the particle picture, the spin of the electrons turned each atom into a small magnet, so that if a shower of atoms is passed between the poles of a fixed magnet, the atoms ought to be affected differently, according to the directions in which the axes of the spin were pointing, and the shower of parallel-moving atoms would be spread out into a broad continuous band. The wave picture, on the other hand, predicts that the shower would be divided into two quite distinct showers, corresponding to the two directions of polarisation of the electron waves. The experiments quite definitely confirmed the predictions of wave-theory.

The failure of the particle picture in this and similar cases is of great interest. For the particle picture implies the possibility, and the wave picture the impossibility, of representation in space and time. So long as we were concerned only with the simplest constituents of nature, electrons, protons and photons, the two pictures appeared to possess equal validity. As soon as we pass to the more complex structure of the atom, the wave picture acquires a definite pre-eminence. Thus the wave picture begins to appear as the true picture of reality, and the particle picture merely as a clumsy approximation to the truth, an approximation obtained by trying to force into a framework of space and time a structure which does not admit of representation in space and time.

This implies that our interpretation of the wave picture as a diagram of the probabilities of finding particles at various spots can no longer be regarded as final. For obviously the true picture of nature must admit of a direct interpretation, without reference to a less perfect picture. The particle representation has served its purpose when it has led us to the wave picture, and may henceforth be disregarded as mere scaffolding.

Thus it is through the wave picture of matter that we must approach reality, and the abandonment of a space-time representation of nature would seem to be the first step on the journey.

It is a difficult step to take. Our knowledge of the external world is brought us by photons which travel in a setting of space and time, with the result that from our earliest days we have thought of objective nature itself as also existing in space and time. Our thoughts have become space-time bound, and can get no grip on concepts outside space and time. Thus no progress has been made along the new road as yet, and we are still compelled to discuss nature in terms of the partial pictures of waves (incomprehensible) and of particles (inaccurate).

Determinism

As the new theory of quanta and the theory of wave-mechanics are believed to agree exactly with observation, and so perhaps to contain the final mathematical truth about nature, they ought to be capable of throwing some light on the question of determinism.

We have seen how our knowledge about nature can be visualised, part by part, in a number of different pictures, although no single picture enables us to visualise the whole truth at once. Of these partial pictures, there are for instance the picture which depicts electrons as particles and that which depicts them as waves. There are again the corresponding pictures for light, one depicting it as waves and the other as photons.

Let us consider the wave picture first. This assumes its simplest form in the case of radiation. The wave-equations become the ordinary equations of Maxwell for the propagation of electric action, and are to all appearances completely deterministic. That is to say, if we know electric conditions at any one instant, we are able, through these equations, to determine these conditions throughout all future time. The wave-equation of an electron implies an exactly similar determinism. If we know the value of ψ throughout space at any one instant, these equations enable us to calculate its value through all subsequent time.

Yet this does not mean that nature is completely deterministic, since, on the only interpretation we have yet been able to devise, both the ψ of Schrödinger's equation and the electric forces of the Maxwell equations are not determined by nature but by our knowledge of nature. If the distinction between nature and our knowledge of nature were to disappear in any special case, as it does, for instance, when we are discussing an assemblage of an immense number of photons, then of course the wave-equations would shew that there was complete determinism in respect of the special phenomenon in question. Other cases of this kind are the two final steady states we have just discussed, but in these our knowledge is not perfect but *nil*. In both cases there is unmistakable determinism, but it is of a very trivial kind. Objectively it is expressed in the sentence: "When a piece of the universe can change no more, its future course is unalterably determined"; subjectively by the equally useless sentence: "If we begin by knowing nothing, and perform no new experiments, we shall continue to know nothing throughout all time".

It is often overlooked that the wholly deterministic wave-equation does not, and cannot, take the whole of nature for its province. Although this is hard to realise, wave-mechanics has no more knowledge of the existence of separate atoms than the undulatory theory of light has of the existence of separate photons. The original analysis of Heisenberg, let us recollect, was not concerned with a succession of photons emitted from a number of distinct atoms, but with a stream of radiation of origin unknown and unspecified. Out of this emerged entities p and q which were made analogous to the momentum and co-ordinate of an electron in an atom by the crude device of adjusting a constant so as to secure agreement when the atom was of very large radius. But the atom in which this electron moved was no real atom such as could exist in nature; it was rather a sort of statistical atom—a composite photograph of all the atoms in the world which conformed to certain specified conditions—as much an abstraction as the "economic man" of the political economist. Wave-mechanics tries to give a concrete picture of this atom, but inevitably this is still a composite photograph, and because of this it extends over the whole plate and is very blurred.

Just because the wave-mechanics deals only with probabilities and statistical assemblies, its apparent determinism may be only another way of expressing the law of averages. The determinism may be of a purely statistical kind, like that relied on by an Insurance Company, or the Bank at Monte Carlo.

This being so, there is no assignable reason why the apparent determinism of the wave-equation should not conceal a complete objective indeterminism. In the mathematical problem known as the "random walk", we imagine that a traveller walks 20 miles a day, but with no causal relation between the directions of his walks on successive days—we can, for instance, imagine his throwing a stick up in the air at random every morning, and letting the direction of its fall determine the direction of his walk for the day. A mathematical formula can of course be obtained to exhibit the chances of his being at various points at successive nightfalls. If we now reduce the unit of time from a day to a second, so that his every step is indeterminate, we find that the probabilities spread out in

traveller's stick fell was not really undetermined, but depended on the force of his throw, which in turn depended on whether he was feeling vigorous, and this in turn on whether the journey of the day before had lain through easy or fatiguing country, and so on indefinitely. If we are to introduce such considerations into our description of nature, it will perhaps take some such form as the following.

We set out to build a conjectural picture of the external world, the only rule of the game being that this picture is to account for our sense-impressions, exactly and down to the smallest detail, and yet is to be objective in the sense of not explaining merely the sense-impressions of a single individual. Each sense-impression is caused by a transfer of energy from the external world to the nerve terminals of our bodies. This transfer is invariably by photons, which, in the new science as in the old, can be adequately represented as travelling in space and time. Thus we naturally begin our conjectural picture of nature by constructing a mental framework of space and time, against which to draw our picture. Going one stage farther, we find that the photons which cause our sense-impressions originate in events. We now find that if our picture is to be objective in the sense just explained, these cannot be represented as localised in space and time separately, but they can still be localised in the blend of space and time we describe as space-time. So long, then, as we do not insist on dividing space-time up into space and time separately, the framework remains adequate for the picture. But these events are the interactions of material objects, electrons and protons and their combinations, and we find that these cannot be adequately depicted as existing in space and time.

Thus our space-time framework proves inadequate for the representation of the whole of nature; it is suited to form a framework for but little more than our sense-impressions, which is precisely the purpose it was originally constructed to serve. We are thus led conjecturally to think of space and time as a sort of outer surface of nature, like the surface of a deep flowing stream. The events which affect our senses are like ripples on the surface of this stream, but their origins—the material objects—throw roots deep down into the stream. When we say a brick is three-dimensional we mean merely that we can only establish contact with it, through our senses, in three dimensions of space. Ripples come from the brick to us in three dimensions of space, but this in no way limits the real existence of the brick to these three dimensions.

Two surface-ripples may appear exactly similar, and may yet be caused by very different happenings down in the depths of the stream, so that the similarity of their appearance provides no guarantee that they will behave in the same way. For this reason we cannot expect the ripple-phenomena on the surface of the stream to shew a strict determinism, nor to conform otherwise than statistically to the law which we describe as the "uniformity of nature". The fact that the surface-phenomena of space-time shew a want of determinism leaves the question of whether real objective nature is deterministic or not completely open.

Space-time is not the framework of the world of nature, but of the world of our sense-perceptions, and when we represent objects beyond our senses in space-time, their apparent absence of determinism may be merely the price we pay for trying to force the real world of nature into too cramped a framework. So, when birds fly through the air, their shadows on the ground beneath obey no uniform or deterministic laws, even though the actual flights of the birds may do so.

If we accept this interpretation, we must conclude that materialistic science runs counter to the teachings of present-day physics in its assumption that everything can be fully represented in space and time; it fails to distinguish between the surface and the depths beneath. It takes the spatial qualities of objects to be their primary qualities, although science shews that the spatial qualities are merely those with which our senses can establish direct contact— the ripples on the surface which meet our eyes.

The purely mechanical picture of visible nature fails for a similar reason. It proclaims that the ripples themselves direct the workings of the universe instead of being mere symptoms of occurrences below; in brief, it makes the mistake of thinking that the weather-vane determines the direction from which the wind shall blow, or that the thermometer keeps the room hot.

CHAPTER VIII

EVENTS

Thermodynamics

The province of atomic physics is to discuss the nature of particular events, and it has been very successful in shewing us how it is that certain kinds of events occur, while others do not. Yet this can give us but little information as to what is happening to the universe as a whole. Another branch of physics, known as thermodynamics, takes this problem in hand; it does not concern itself with individual events separately, but studies events in crowds, statistically. Its province is to discuss the general trend of events, with a view to predicting how the universe as a whole will change with the passage of time.

The science of thermodynamics had its origin in severely practical problems relating to the efficiency of engines, but it was soon extended to cover the operations of nature as a whole. All this happened in the days when nature was assumed, without question, to be mechanical and deterministic. In what follows, we shall not treat nature as mechanical, but for the moment we shall treat it as though it were strictly deterministic.

On a deterministic view of nature, the universe never has any choice; its final state is inherent in its present state, just as this present state was inherent in its state at its creation. It must inevitably move along a single road to a predestined end, like a train rolling along a single-track line, on which there are no junctions of any kind. Thus if a super-experimentalist could discover the exact position and the exact speed of motion of every particle in the universe at any single instant, a super-mathematician would be able to deduce the whole past and the whole future of the universe from these data.

Experimental physics has not yet been able to provide such data, and the uncertainty principle shews that it never will be. Yet a super-mathematician, who had un-limited time at his disposal, might calculate out all the different pasts and futures which would result from all conceivable sets of data—in other words from all con-ceivable present states.

He might commence his labours by making a diagram in which to map out all possible states of the universe, just as all points in England are mapped out in an ordinary geographical map. He could start from any particular point in this diagram and trace out, by mathe-matical calculation, the whole future of a universe which started from the state represented by this point. He could represent this future by a line through the point, which would run through his diagram much as a railway line is represented by a line running across the map of England. He could take point after point in his diagram in turn, and represent the development of a universe which started from each point by a line, until his whole diagram was filled with lines. These lines would represent all the lines of development which were possible for the universe. If the universe was strictly deterministic, as we have so far supposed, the diagram would look like the map of a country covered with single-track lines of railway, with no junctions of any kind. If, on the other hand, strict determinism does not prevail in the universe, there may be any number of junctions and connecting tracks between the different lines.

Let us imagine that a perfect diagram of this kind is at our disposal, as it would be—in theory at least—if we had a perfect knowledge of the laws of nature. No matter how perfect the diagram is, we are still unable to gain a detailed knowledge of our future from it, because we do not know our present position on the map. This makes it impossible to identify the particular track on which we are travelling, so that we can neither say what part of the diagram it will traverse next nor where it will end. Yet it may be possible to discover in what kind of country it ends, and this is the information we really want. It is information of this kind that the science of thermodynamics can provide.

Imagine that we suddenly waken up from a state of unconsciousness to discover we are on a British railway. We have no means of knowing where our journey will end. Yet if we have a physical map of Great Britain with us, we may notice that only a few hundred acres out of 55 million lie as much as 4000 feet above sea-level. Although we cannot say where our journey will end, there are obviously very long odds that it will end at a height of something less than 4000 feet above sea-level. If a barometer in our compartment indicates that we are already as much as 4000 feet above sea-level, then there are very long odds that the general trend of our journey will be downhill.

It is to considerations of this kind, rather than to exact knowledge, that we must turn for guidance in our efforts to study the evolution and final end of the universe. As certain knowledge is beyond our reach, we must be guided entirely by probabilities. Yet the odds we encounter in calculating these probabilities prove always to be so immense that we may, for all practical purposes, treat long odds as certainties. Because the number of particles— electrons and protons—in the universe is of the order of 10^{79}, we find that high powers of 10^{79} enter into all our odds, and, this being so, we need not trouble to differentiate too carefully between probabilities of such a kind and certainties.

Entropy

Thermodynamics is much concerned with a quantity known as "entropy". This plays much the same part in our diagram of the universe as height played in our imaginary railway map of Great Britain, except that *small* entropy corresponds to *great* height, and vice-versa; thus entropy does not correspond so much to height above the level of the sea, as to depth below the top of the highest mountain. The highest mountain in Great Britain rises to 4400 ft. above sea-level, and as most of Great Britain is only a little above sea-level, most of it is at a depth, in this sense of the word, of nearly 4400 ft.—the maximum depth possible. In the same way, we find that most of the configurations which figure in our map of the universe are at the maximum entropy possible—all, indeed, except for minute regions whose sizes are proportional to inverse powers of 10^{79}.

At the moment we cannot justify this statement because we have not yet defined "entropy". And there is no need to justify it, because the best definition of "entropy" makes the statement true of itself and automatically. It is convenient to define "maximum entropy" as specifying the condition which is commonest in our map of the universe, and then, having done this, to define entropy in general in such a way that the more common condition is always of higher entropy than the less common. Thus we define entropy to be a measure of the "commonness" of a given state in our map.*

With this definition we find that, just because the numerical factors involved are so immense, conditions of "maximum" entropy are not only more common, but *incomparably* more common, than those whose entropy is less, and so it is all down the ladder. Because of this, it is practically certain that each state of the universe will be succeeded by a state of higher entropy than itself, so that the universe will "evolve" through a succession of states of ever-increasing entropy, until it finally reaches a state of maximum entropy. Beyond this it cannot go; it must come to rest—not in the sense that every atom in it will have come to rest (for maximum entropy does not involve this), but rather in the sense that its general characteristics cannot change any more.

Yet if someone asserts that this will not happen, and that the universe will move to a state of *lower* entropy than the present, we cannot prove him wrong. He is entitled to his opinion, either as a speculation or as a pious hope. All we can say is that the odds against his dream coming true involve a very high power of 10^{79} in his disfavour.

Thermodynamics is accustomed to disregard all such

* If W is the "commonness" of a certain state, the mathematician defines the entropy of this state as $k \log W$, where k is the gas-constant.

infinitesimal chances and forlorn hopes, and announces its laws as certainties. We must nevertheless always bear in mind that there is a small risk of failure attached to every such law. The famous "second law of thermodynamics" asserts that the entropy of a natural system always increases, until a final state is attained in which the entropy can increase no further; a fuller statement of the law would be that the chances of the entropy doing otherwise are negligibly small.

The Final State of Maximum Entropy

We now see that the question of discovering the final state of the universe is merely that of discovering how far the entropy of the universe can increase without violating the physical laws which govern the motions of its smallest parts. There was no need to take the physical properties of matter into account in defining entropy, but we must do so before we can discover the state in which the entropy is a maximum.

The process is usually very complicated, but two simple instances may illustrate the general characteristics of a state of maximum entropy. They do not refer to the universe as a whole, but merely to minute portions which have been selected for their simplicity and familiarity.

Let us pour some red ink into water, and leave the ink and the water to diffuse into one another. We know, before the event occurs, that the final state will be one in which they are uniformly mixed to form a homogeneous pinkish fluid, and as this state of uniform mixture is invariably the final state, we know that it must be the state of maximum entropy.

Again, let us put a kettle of cold water over a hot fire. We know, before we perform the experiment, that the final state will be one in which all the water is turned into steam. This also must be a state of maximum entropy. Just as the red ink diffused itself equally through all parts of the water in attaining a state of maximum entropy, so the heat of the fire tends to diffuse itself equally through coal, kettle, and water.

These instances have shewn us two final states in which the entropy is a maximum. They illustrate a very wide and very general principle—the final state of maximum entropy avoids concentration, whether of special substances (as with the ink) or of energy (as with the heat of the fire). The "commonest state" is one in which both substance and energy are uniformly diffused, just as the commonest state in which we find a concert audience is that in which tall people and short, dark and fair, and so on, are uniformly diffused.

General considerations of this kind can tell us something at least as to the final end of the universe, but they cannot indicate the road by which it will be reached. All they can tell us is that the road is practically certain to be one of increasing entropy throughout; and the better we understand entropy, the more this statement will convey to us. It is not impossible for the entropy to decrease, but it is almost infinitely improbable that it should do so.

For instance, when the ink and water have once become thoroughly mixed, the state of maximum entropy has been attained; the ink-water mixture cannot change its general characteristics any further without a decrease of entropy. Yet the molecules of ink and water still jostle one another about, and change places as they do so. It is quite conceivable that their random motions should take them into a configuration in which all the ink molecules are found at one end of the vessel, and all the water molecules at the other. The entropy of such a configuration is far below the maximum possible, so that the odds against the molecules of ink and water assuming such a configuration are immense. Yet it is important to notice that no law of nature prohibits it. Indeed, if we had an infinite number of vessels of ink and water, the unexpected would be bound to happen in a few of them—just as, if an enormous number of hands of bridge are played, there are bound to be a few deals in which each player gets one complete suit, in spite of the immense *a priori* odds against such an event occurring in a single individual case. The event is bound to occur either if an enormous number of players play bridge for a short time or if a single party of players play for an enormous time. In the same way we may say that a complete separation of the ink and water is bound to occur, either if we have an infinite number of vessels containing the mixture, or if a single vessel exists for an infinite time.

Similar considerations apply to our other miniature universe of fire, kettle and water; the water in the kettle may freeze as the result of being put over a hot fire. To prove this, we need only notice that there is a possible state of this group of objects in which the water exists in the form of ice, and the fire is even hotter than before because there is less heat in the kettle and its contents. If we map out all the configurations of the system, this particular configuration must appear on the map, so that we cannot know for certain that it will not be the end of the journey. We know, however, that when we put a kettle of ice on the fire the normal event is for it to turn into a kettle of water. This shews that the entropy of the water-configuration is higher than that of the ice-configuration, and this in turn shews that although it is possible for a kettle of water to freeze when placed over a hot fire, it is almost infinitely improbable that it will do so on any single occasion. If even the most credible of witnesses told us a kettle of water had frozen when he put it on a hot fire, we should not believe him, although there is nothing in the laws of nature to prohibit such an occurrence; indeed these very laws assure us that the event must occasionally happen. Yet such occasions must from the nature of things be so very rare, that we should think it far more likely that our informant had gone crazy, had been deceived, or was lying, than that he had been present at one of them.

These examples have both illustrated cases in which the individual atoms and molecules are left to perform random motions under the play of blind forces. If the atoms and molecules receive any kind of guidance, the result may be very different. Suppose that, instead of ink, we pour oil into our water. We no longer expect the final result to be a uniform mixture; we know that we shall find all the oil on top and all the water below. An arrangement which is inconceivably improbable for ink and water is found to be the most probable of all for oil and water; indeed, exact calculation confirms that a state of practically complete separation is the state of maximum entropy in the case now under consideration. The reason for the change is that the force of gravity differentiates between the molecules of oil and of water. When we say that oil is of lower specific gravity than water, we mean in effect that the earth's attraction draws particles of water downward with a force greater than it exerts on equal-sized particles of oil. Because it continually drags these latter particles down with a smaller force, it encourages them to move upwards through the water. When we mix oil and water, we are not handing over their molecules to be the play-things of a blind chance, but rather to a chance over-ridden by the selective action of gravitation. There is blind interplay of the molecules of oil between themselves and of the molecules of water between themselves, but the cross interplay is controlled by gravitation.

Suppose, for instance, that we divide our vessel into two equal divisions, each holding a pint, by a horizontal membrane with a small pinhole in it. Let us mix a pint of oil and a pint of water as thoroughly as possible, and fill our vessel on both sides of the membrane with this quart of mixed liquid. After a sufficient time, we shall of course find that all the oil has passed into the upper half, while all the water has passed into the lower half; our careful mixing has been undone, and this by very simple means. Whenever a particle of oil in the lower half met a particle of water in the upper half at the pinhole— the only place at which they could meet—the force of gravity urged them to change places, and such inter-changes have continually increased the amount of oil in the upper half and that of water in the lower half, until complete separation has been effected.

The Sorting Demon of Maxwell

If we performed a similar experiment with our previous mixture of water and red ink, no such action would take place in the ordinary course of nature, since gravity makes no distinction between liquids of the same specific gravity. Yet suppose an intelligent being of microscopic size were placed at the pinhole, armed with a tiny shutter with which he could close the aperture when he wished, and was given instructions to open it only for molecules of ink passing upwards or for molecules of water passing downwards—in brief his task would be to perform a selective action like that which gravity performs for oil and water. It is clear that after a long enough time the ink and water will be as thoroughly separated as the oil and water had previously been, although this time the separation would have been produced not by gravity but by intelligence.

The intelligent microscopic being we have just described was introduced into science by the Cambridge physicist Clerk Maxwell, and is generally described as "Maxwell's demon". The demon, we must notice, in no way sets himself in opposition to the laws of mechanics. We do not know how often he finds it necessary to open and close his microscopic shutter. The natural motions of the molecules may conceivably be such that he finds no occasion to close it at all. Then everything will go on precisely as though the demon had not been called on to help, the ink and water separating out under their own natural random motions. Yet the odds against such an occurrence are unthinkably large. As each separate molecule comes into view, the demon must ask himself the question "To act or not to act?", and then put his decision into practice. A prolonged run of decisions all in the same sense will be as improbable as a prolonged run of heads or of tails when we spin a coin. Thus it is exceedingly unlikely that our demon will find that no action is needed time after time; the normal event will be that he will need to open and close his shutter millions of times. Even so, he expends no energy in so doing, and each time the shutter is closed against a molecule, we may reflect that had the path of the molecule in question been a hair's-breadth to right or left, it would have bounced off the membrane without the demon touching his shutter.

Although the demon does not interfere with the operation of the laws of nature, yet he exercises a selective effect, and by this alone he can cause any system to pass to a state of lower entropy. Natural forces, left to their own blind interplay, are practically certain to increase the entropy, but it is the play of the laws of probability rather than of the laws of nature that produces this result. The demon has not been told to circumvent the laws of nature, but the laws of probability; he can so to speak load the dice from moment to moment, and obtain any result he wants provided this does not violate the laws of nature—the conservation of mass, of energy, and so forth. When red ink and water are mixed, he cannot increase the total amount of either or both; all he can do is to disentangle them, as one might sort out a heap of red and white beads, or again as a railway shunter divides up a goods train by moving the switches in different ways for different wagons. When a kettle of water is placed over a fire, he cannot add to the total amount of heat, but he can, if he wishes, increase the heat of the fire by subtracting heat from the kettle. His accomplishments are limited to robbing Peter to pay Paul, whereas unaided nature would leave Peter and Paul to fight it out—or perhaps to toss up for it time after time.

Quite general considerations shew that the universe as a whole has a very long way to go before coming anywhere near its final state of maximum entropy. In this final state, concentrations of radiation and of temperature will equally have disappeared, so that radiation will be distributed uniformly throughout space, and the temperature will be everywhere the same. At present, the density of radiant energy out in the farthest depths of space corresponds to a temperature of less than one degree above absolute zero; in the interstellar spaces of the galactic system, to three or four degrees only; near the earth's orbit to about 280 degrees; at the sun's surface to about 6000 degrees; at the sun's centre to perhaps 40 or 50 million degrees. The universe can always increase its entropy by equalising these temperatures; as for instance by letting energy flow from the sun's hot centre to its cooler surface, by letting it then stream out into space, past the earth's orbit, into the cold and dark of interstellar and intergalactic space. There can be no end to the increase of entropy until these regions are all at the same temperature, with radiant energy diffused uniformly throughout space. Then, and then only, will the universe have reached its final state, a state in which the temperature will everywhere have fallen too low for life to exist—the perfect quiet and perfect darkness of eternal night.

The Activities of Life

A general survey of the universe as a whole suggests that it is rapidly moving towards such an end. The sun is dying, pouring out some 250 million tons of its substance in the form of radiation each minute, thereby lowering its own heat and raising that of empty space. Other stars tell the same story; we find no evidence of sorting demons sitting on their surfaces to turn the heat back into their hot interiors. Yet a being from another universe who scrutinised this earth of ours with sufficient care might notice signs which led him to wonder whether there might not be local exceptions to the general increase of entropy. For instance, regarded from the purely physical point of view, gold is a fairly ordinary metal; natural laws shew it no favour nor special treatment. Yet our visitor might notice that the world's total supply of gold, which had originally been fairly uniformly scattered throughout parts of the earth's crust, tended to become highly concentrated in a few small regions, in a way which would seem to set the demands of the second law of thermodynamics utterly at defiance. Again, the law does not approve of fires occurring at all, although it admits that accidents will happen. It insists, however, that these accidental violations are most likely to occur when the weather is hot and dry. Yet our observer would not only detect innumerable fires on earth, but would notice that they occurred most frequently when it was cold and damp; he would see more in those parts of the surface of the earth which were covered with winter snow than in those which were parched with equatorial or summer heat. On the other hand, he might notice that small accumulations of ice were especially in evidence when the weather was hot and sultry.

The odds against all these events occurring in the normal course of a nature which had not been tampered with would be of the same order as the odds against a kettle of water freezing when placed on a hot fire. Thus no picture of nature can claim to be complete, unless it contains some means by which the statistical laws of nature may be evaded—if not throughout the whole of nature, at least in chosen spots on our own earth. Our visitor might perhaps conjecturally attribute these evasions to the activities of innumerable sorting demons.

A statistical survey of the more violent offences committed against the second law of thermodynamics would shew that the hotbeds of crime are precisely those places we describe as centres of civilisation. Inanimate matter obeys the law implicitly; what we describe as life succeeds in evading it in varying degrees. In fact it would seem reasonable to define life as being characterised by a capacity for evading this law. It probably cannot evade the laws of atomic physics, which are believed to apply as much to the atoms of a brain as to the atoms of a brick, but it seems able to evade the statistical laws of probability. The higher the type of life, the greater is its capacity for evasion. And the observed evasions so closely resemble the results that would be produced by an army of sorting demons, that it would seem permissible to conjecture that life operates in some similar way.

So long as nature was believed to be mechanistic, and therefore deterministic, such a conjecture was hardly permissible—the sorting demons would have interfered with the predestined course of nature.

On the other hand, modern physics can adduce no such objection to the conjecture; the only determinism of which it is at all sure is of a merely statistical kind. We still see the actions of vast crowds of molecules or particles conforming to determinism—this is of course the determinism we observe in our everyday life, the basis of the so-called law of the uniformity of nature. But no determinism has so far been discovered in the motions of the separate individuals; on the contrary, the phenomena of radio-activity and radiation rather suggest that these do not move as they are pushed and pulled by inexorable forces; so long as we picture them in time and space, their future appears to be undetermined and uncertain at every step. They may go one way or another if nothing intervenes to direct their paths; they are not controlled by pre-determined forces, but only by the statistical laws of probability. If an unknown something intervenes to guide them, they may transfer their allegiance from the laws of probability to the guiding something, as the molecules of the oil-water mixture did to the force of gravitation. There seems no longer to be any reason why this something should not be similar to the action of sorting demons, the volitions of intelligent minds loading the dice in their own favour, and so influencing, so to speak, the motions of the molecules when they are in doubt which path to take— provided always that volitions and molecules are not too dissimilar in their nature for such interaction to be possible.

Space-time and Nature

We can also look at the matter in the alternative way described on p. 258. We have just been picturing nature as an assemblage of particles set in a framework of space and time. Yet we have seen elsewhere that such a framework is not suited for the arrangement of the whole external world, but only for the photons by which it sends messages to our senses. Because these messages arrive in a framework of space-time, we must not conclude that the whole external world exists within the confines of the same framework. Our observational knowledge of the outer world is limited by the aperture of our senses, and these form blinkers which prevent our seeing beyond space and time—just as our telescope may prevent us seeing more than a small angle of the sky. But the events we see in space and time may have their origin outside space and time—just as the curve in the tail of the brilliant comet we see in our telescope may have its origin in the sun which lies outside the field of the telescope. The recent developments of theoretical physics suggest that this may be the case with many of the phenomena of physics. It has proved impossible to find any description of electrons and protons in space and time such as shall fully account for the phenomena originating in them.

This has led us to think of space-time as a sort of surface-layer of the universe; the sources of events appear in this space-time surface in the form of material protons and electrons, but they have their roots in a deeper stratum. Thus although no causality may be discernible while we limit our vision to the surface of things, yet if we

could take the whole of reality into view, we might see cause
and effect inter-related, events following clearly specified
laws, and not occurring merely as illustrations of the laws
of probability. The gardener plants a dozen trees which,
so far as he can see, are exactly similar; he has been told
that only fifty per cent. of trees of this species thrive, and
this is confirmed when he finds that, out of his dozen trees,
six thrive and six fail. Yet he does not attribute their
different fates to the laws of probability, but to happenings in the soil beyond his vision. He digs down and
finds wire-worms at the roots of the six failures. The wire-
worms play much the same sort of part as we have
imagined the sorting demon to play in our space-time
picture, or as we have conjectured that our volitions and
intelligences might play. Residing beyond the stratum of
time and space, they can influence events, which also have
their roots outside time and space, and so exercise some
control over the happenings in time and space.

Conjectures of the kind bring us to a region of thought
in which human predilections are deeply concerned. Some
who are eager to find a place for virtue, beauty and other
"values" in the scheme of things are very ready to hail any
evidence of indeterminism in nature as almost affording
a proof of human free-will. Others refuse to admit the
possibility of indeterminism even in nature, and insist that
we, like all nature, are mere cogs of a machine which is
running down to its predetermined end.

Apart from extremists, a number of moderate men still
adopt an attitude of extreme caution, and even suspicion,
towards any attempt to reconcile human free-will with
the scheme of physical science. Many quote recent investi-

gations in physiology and psychology as providing evidence, not against the possibility of free-will, but against its probability. Others regard the present situation in physics as a mere transitory phase. For instance, Planck, who has given much thought to this question, writes, with reference to the impact of quantum ideas on the fundamental laws of physics:[*]

"Some essential modification seems to be inevitable; but I firmly believe, in company with most physicists, that the quantum hypothesis will eventually find its exact expression in certain equations which will be a more exact formulation of the law of causality",

and is prepared to extend the operation of this law to human activities:

"The principle of causality must be held to extend even to the highest achievements of the human soul. We must admit that the mind of each one of our greatest geniuses— Aristotle, Kant or Leonardo, Goethe or Beethoven, Dante or Shakespeare—even at the moment of its highest flights of thought or in the most profound inner workings of the soul, was subject to the causal fiat and was an instrument in the hands of an almighty law which governs the world".

Einstein is reported as expressing similar opinions:[†]

"I am entirely in agreement with our friend Planck in regard to the stand which he has taken on this principle. He admits the impossibility of applying the causal principle to the inner processes of atomic physics under the present state of affairs; but he has set himself definitely against the thesis that from this *Unbrauchbarkeit* or inapplicability we are to conclude that the process of causation does not exist in external reality. Planck has really not taken up

[*] *Where is Science going?* by Max Planck (1933), pp. 143, 155.
[†] *L.c.* pp. 210, 201.

any definite standpoint here. He has only contradicted the
emphatic assertions of some quantum theorists and I agree
fully with him. And when you mention people who speak
of such a thing as free will in nature it is difficult for me to
find a suitable reply. The idea is of course preposterous.. . .

"Honestly I cannot understand what people mean when
they talk about freedom of the human will".

Weyl, on the other hand, after explaining how the limits
to determinism, if any, will be found by passing along the
road from the large scale phenomena of astronomy and
physics, which necessarily appear deterministic (p. 230), to
the small scale phenomena at the far end of the road,
continues*:

"We firmly believe today that we have touched these
limits in quantum mechanics....

"At the same time 'fate' as expressed in the natural laws
appears to be so weakened by our analysis that only through
misunderstanding can it be placed in opposition to free
will".

I do not think that either the facts of physical science or
their interpretation within the legitimate province of physical science are in dispute among men of science; on the
contrary, I believe we are all in agreement. Differences
only arise when physicists take to speculation either about
the future progress of science (as Planck does in the above
quotation), or about the ultimate problem of human free-
will, which of course lies beyond the province of physics.
The famous dictum of Schopenhauer—"Man can do what
he wills, but cannot will what he wills"—contains two
distinct statements. The latter has to do with happenings
on the mind side of the mind-body bridge, which are not

* *The Open World*, pp. 35, 43.

the concern of physics. The former is the concern of physics. In brief, it was believed to be in conflict with nineteenth-century physics, but is not in conflict with the physics of to-day; whether it will be in conflict with the physics of to-morrow remains to be seen.

Nevertheless, the most we can say is that crevices have begun to appear in what used to be considered the impregnable closed cycle of physical science. Whether the volitions of the human mind can pass through these and affect the operations of nature must in the last resort depend on whether the two are sufficiently alike to interact—a keyhole is useless unless we have a key of the same nature as the lock. It may still be, as Descartes maintained, that mind is too dissimilar from matter ever to be able to influence it.

Mind and Matter

A century after Descartes, we find Berkeley maintaining that we had no right to say that matter was different from mind. With no knowledge of matter except such as comes to us through the perceptions of our minds, what warrant can there possibly be for supposing the two are of unlike natures? Matter outside our minds produces ideas inside our minds; causes must be of like nature to their effects, and "after all, there is nothing like an idea but an idea". Thus Berkeley argued that matter must be of the same general nature as an idea, like the matter we see in a dream. To say that mind cannot influence matter now becomes as absurd as to say that mind cannot influence ideas.

A later school of philosophy shewed how this argument

could be turned against its author. Even if matter and
mind were of similar nature, how did we know they were
of the nature of mind rather than of matter? The science
of that time claimed to know a great deal about matter,
but admittedly knew very little about mind; thus it was said
that the scientific picture of matter must also portray mind
and its operations. And this picture was that to which
we have so often referred—a jumble of mechanical atoms
moving blindly along their pre-arranged paths to predestined ends.

The logic of this argument stands, but not the premised
picture of matter. In so far as science now draws any
picture at all of matter, it is one which seems in every
way closer in mind.

To some extent this must be the case. The old science
which pictured nature as a crowd of blindly wandering
atoms, claimed that it was depicting a completely objective universe, entirely outside of, and detached from, the
mind which perceived it. Modern science makes no such
claim, frankly admitting that its subject of study is primarily
our observation of nature, and not nature itself. The new
picture of nature must then inevitably involve mind as
well as matter—the mind which perceives and the matter
which is perceived—and so must be more mental in
character than the fallacious picture which preceded it.

Yet the essence of the present situation in physics is not
that something mental has come into the new picture of
nature, so much as that nothing non-mental has survived
from the old picture. As we have watched the gradual
metamorphosis of the old picture into the new, we have
not seen the addition of mind to matter so much as the

complete disappearance of matter, at least of the kind out of which the older physics constructed its objective universe.

The Einstein-Heisenberg policy of concentrating on observables might well have been adopted in the first instance as a mere matter of scientific technique; it was an obvious precaution to make as few assumptions as possible about unobservables, and so lessen the risk of unjustified assumptions and wasted work.

Such a policy might have resulted in either of two ways. If the phenomena of observation were evidence of an objective nature existing in its own right, the procedure might have been expected first to co-ordinate the observables and then to throw some light on the real nature of the unobservables behind them. If, on the other hand, nature was largely or wholly subjective, the procedure might have been expected to disclose this fact. Actually the result has come nearer to the latter alternative than to the former. The observables do not appear to owe their existence to our supposed unobservables existing in the reality behind them so much as to our conscious minds observing them from in front. Electrons, protons and their varied arrangements seem as unable to provide true primary qualities as were the older mass, motion and extension in space of Locke and Descartes; the theory of quanta seems to dethrone the former as effectively as the theory of relativity dethroned the latter. Thus the procedure of concentrating on observables appears to be leading to results different from those which might have been anticipated if the unobservables had existed in their own right; it seems to lend a new meaning to the dictum "Esse est percipi" of the philosophy of an earlier age.

Such considerations as these undoubtedly introduce a
markedly subjective tinge into all discussion of the present
situation in mathematical physics. We must, however, be
on our guard against taking a wrong turning at this point
of our discussion. Even if our assumed unobservables—
electrons and protons—should prove to be wholly subjective,
this would not prove that all nature is subjective. Our
unobservables are at best mere guesses. These particular
unobservables may have been bad guesses, mere creations
of our own imagining, but this does not shew that others
might not have been good guesses. Perhaps the proper
interpretation of the situation is merely that we must look
for new unobservables.

It is not difficult to know where to look. We have already
seen that the particle picture, which treats matter as consisting of electrons and protons, fails, in some respects, to
represent the true properties of matter; the wave picture,
on the other hand, is nowhere known to fail, and so may
provide the true gateway to reality (p. 253). Now the
waves of this picture are of course unobservables; it may
be that a study of these, rather than of electrons and
protons, will lead us to the true objective reality behind
appearances. Our attempt to relate these waves to particles
introduced subjectivity, but this may have entered from
the particle side and not from the wave side of the attempted
relation. We have so far interpreted waves as specifying the
probabilities of particles existing at points in space-time;
we may equally well interpret them as specifying the
probabilities of happenings at points in space-time—the
spot of light on the screen, the blackening of the photographic plate by the impact of the supposed "electron".

Indeed, if these waves are to lead us to an objective reality, we must associate them with happenings rather than with particles, since we have already seen (pp. 200, 250) that they have no objective existence for particles to which nothing is happening beyond bare existence. The quantum theory seizes the photons from a source of radiation at the moment of their emergence into space-time, analyses them and tries to refer them to the motion of an assumed electron (p. 184) under the electric attraction of a nucleus. Yet this procedure leads to no objective specification for the assumed electron when it is away from the electric field; we found that we could only make our picture of matter objective by leaving isolated electrons and protons out of it altogether; these seem to acquire objective reality only when combined to form an atom, and so to produce events (p. 249), just as our individual space and time are found only to acquire objective reality when they are combined to form a four-dimensional space-time.

It may be objected that, as nothing is put into the theory except our knowledge of radiation, we can hardly expect a positive knowledge of objects to emerge. Yet if the electron and proton had permitted of separate objective specifications, we might have expected to be able to distinguish the two ingredients separately in the specification of the combination. It has not proved possible to do this; the quantum theory does not encourage us to regard the combination as the juxtaposition of two particles, but merely as a source of radiation issuing into space-time.

We cannot explain the situation away by saying that the uncertainty principle makes objective specifications impossible; this is putting the cart before the horse. The impossibility of objective specifications is inherent in the wave picture, so that if, as we now suppose, the wave picture is fundamental, the uncertainty principle is the consequence, and not the cause, of this impossibility.

This inevitably raises doubts as to whether the isolated electron and proton have any existence at all in reality. The theory of relativity raises the same doubts, although in a somewhat different form. For, after experimental physics has reduced the supposed matter to its ultimate constituents, electrons and protons, the theory of relativity finds it necessary to carry the resolution further. According to the older physics, a particle of matter was characterised by continued existence in time. The theory of relativity represents this continued existence by a continuous line in space-time, and then resolves this line into its points, each of which represents an "event"—the existence of the particle at a single instant of time. Space-time is warped at every point, and in particular at the points along this line. Yet the warping at these points does not differ in essential character from that elsewhere. If the particle had no extension in space beyond that of a mere point, we might find a sharp edge or ridge of warping, but we have seen that we cannot assign to the elementary particles either a definite localisation or a sharply defined boundary; the wave picture of a particle, whatever else it may be, is never a point. Thus the "world-line" of a particle is, strictly speaking, not a line at all, but is a continuous and unbounded curved region, and must logically be separated into small curved spots—the particle resolves itself into events. Most of these events are unobservable; it is only when two particles meet or come near to one another that we have an observable event which can affect our senses. We have no knowledge of the existence of the particle between times, so that observation only warrants us in regarding its existence as a succession of isolated events.

It may be objected that all nature goes on as though these particles had a real existence, and this provides presumptive evidence that they have. A similar argument might of course be adduced to prove the real existence of photons; we have seen that the evidence for their existence is of the same general type as that for electrons (p. 155), and that a large part of nature can be explained by supposing photons to have a real existence (p. 156). Indeed it is easy to imagine beings in intergalactic space, where matter is rare, endowed with electric senses in place of our material senses, who would regard photons as the primary constituent of reality, and matter as something outside the general course of nature. Yet we have seen that photons are merely combinations of free vibrations, so that if the wave picture is fundamental, photons cannot be said to have a real existence of the kind which we used to attribute to electrons. And, if photons must be dismissed from the realm of reality, it is hard to find any reason for retaining electrons and protons.

It becomes important at this stage to make a clear distinction between existence and identifiable existence. For instance, the pounds, shillings and pence of our bank accounts have a real existence, but not an identifiable existence; we cannot say they are Bank of England notes of numbers so and so. In physics it is the same with energy; it would for instance be meaningless and silly to say that the energy which is now lighting my room is identical with that with which Samson pulled down the pillars of the house in Gaza; energy has no identifiable existence. Again it is the same with electrons. When two dogs *A*, *B* engage in a dog-fight, two damaged dogs *C*, *D* emerge of slightly altered appearance; but it is always possible to say, for instance, that *C* was *A* and *D* was *B*. But when two electrons meet in an encounter, this is not the case; the identification is not only impossible in practise but is meaningless in theory. The mere assumption that it is possible leads to difficulties and wrong results in physics. Yet when electrons and protons are combined to form an atom, this atom appears to retain an identifiable existence, at least through long periods of time. It is not meaningless to say to-day that certain atoms of gold formed part of Cleopatra's crown, but it is meaningless to say that certain electrons formed part of the pearl she drank. No doubt it is often convenient to regard events as strung on to electrons and protons, like beads on a thread, but the manner of stringing is merely a matter of subjective choice; I may string them in one way, and you in another, and both ways are equally valid. Thus the events must be treated as the fundamental objective constituents, and we must no longer think of the universe as consisting of solid pieces of matter which persist in time, and move about in space.

On some such grounds as these it is possible to conjecture, with Leibnitz, that matter as ordinarily understood, the matter of solid objects and hard particles, has no existence in reality, and only appears to exist through our observing non-material things in a confused way—through the bias of our human spectacles. Events and not particles constitute the true objective reality, so that a piece of matter becomes, in Bertrand Russell's words,

"not a persistent thing with varying states, but a system of inter-related events. The old solidity is gone, and with it the characteristics that, to the materialist, made matter seem more real than fleeting thoughts".

This at once takes all force out of the popular objection that mind and matter are so unlike that all interaction is impossible. With matter resolved into events, the objection is no longer tenable. We see the territory on both sides of the mind-body bridge occupied by events, and as Bertrand Russell says:*

"The events that happen in our minds are part of the course of nature, and we do not know that the events which happen elsewhere are of a totally different kind".

There is then no longer any reason, on these grounds, why the two should not interact. This of course brings us to something which is very like Berkeley's famous argument, clad in modern dress, and supported by scientific knowledge. It obviously follows that, to quote Russell again,†

"the world presented for our belief by a philosophy based upon modern science is in many ways less alien to ourselves than the world of matter as conceived in former centuries".

The Mathematical Pattern

Einstein has written:‡

"In every important advance the physicist finds that the fundamental laws are simplified more and more as experimental research advances. He is astonished to notice how

---

* *Outline of Philosophy*, p. 311.　　† *L.c.* p. 311.
‡ Introduction to *Where is Science going?* p. 13.

sublime order emerges from what appeared to be chaos. And this cannot be traced back to the workings of his own mind but is due to a quality that is inherent in the world of perception".

Weyl has made a similar comment, writing: *

"The astonishing thing is not that there exist natural laws, but that the further the analysis proceeds, the finer the details, the finer the elements to which the phenomena are reduced, the simpler—and not the more complicated, as one would originally expect—the fundamental relations become and the more exactly do they describe the actual occurrences."

We have had ample evidence of this tendency toward simplicity in the present book. We have seen Hero's simple synthesis of the two laws of Euclid gradually expanding in scope until it embraces almost all the activities of the universe, and yet maintaining its original simplicity of mathematical form throughout. Phenomenal nature is reduced to an array of events in the four-dimensional continuum, and the arrangement of these events proves to be of an exceedingly simple mathematical kind. The discovery of the pattern underlying the arrangement might have been expected to suggest some reason why this special arrangement prevailed rather than another. It is as though we had set out to study the fundamental texture of a picture and had found this to consist of regularly spaced dots, as in a half-tone print. We are not concerned with the meaning of the picture as a whole, which may be moral or aesthetic or anything else; this is not the province of science. We are concerned only with the fundamental

* *The Open World*, p. 41.

texture of the picture, which might conceivably have told us something as to its physical nature, something for instance as to the substance on which the picture was printed. But science has so far been unable to discover anything about the dots except the exceeding simplicity of their arrangement.

This simplicity is of a mathematical kind; it seems to admit of a very simple mathematical interpretation and of no other, as though, in Boyle's phrase, mathematics is the alphabet of the language in which nature is written. The words of this language may or may not be mental in their meanings; the immediate point is that, even in the alphabet, we can discover no reality different in kind from that we associate with a mere mental concept. These mental concepts are not of the kind that we associate with the work of the engineer or the poet or the moralist, but with the thinker who works with pure thought alone as his raw material, the mathematician at work in his study.

Space provides an obvious instance of this. The concept of a finite space reduces the science of astronomy to law and order, just as the concept of a finite surface for the earth reduces the science of geography to law and order. It is easy to make a model of the earth's surface. We merely take any spherical object, and its surface—the transition from matter to something which is not matter—gives us our model. But we cannot make a model of a finite space in the same way, because we cannot imagine a layer of transition from space to something which is not space. Anyone who mentions the finiteness of space in his writings or lectures is besieged with questions as to what lies beyond the finite space. It is impossible, we are told, to think of finite space as a physical reality. If we try to do so, we are at once asked what is outside the space. What can there be except more space?—and so on *ad infinitum*, which proves that space cannot be finite.

If we give up trying to attach any sort of reality to finite space except that of a purely mental concept, our way immediately becomes clear. Our everyday thoughts are never concerned with more than a finite part of space, so that finite space as a framework for mental processes is familiar to us all.

It seems likely that to bring law and order into the phenomena of nature, we shall further have to suppose that the finite space is expanding, and this raises similar questions. What can space expand into, except more space? Yet if it does so, the space which expands cannot be the whole of space, and so on as before, whence it follows that the whole of space cannot be expanding. Thus we cannot attribute any reality to the space of the universe, except again as a mental concept; any attempt to assign a degree of reality different from this to space leads only to confusion and contradictions.

It may be urged that this does not prove anything new, since we already know that space cannot have any objective reality except as one constituent of the continuum. But similar considerations apply to the continuum itself, the one entity in which science absorbs all others, and to which alone an objective reality seems possible. We find that we must picture this also as limited, so that unless we treat this also as a mere mental concept, we are confronted with the question as to what lies beyond the limits. Yet when we so treat it we find we have reduced the whole of nature to a mental concept, since the texture of nature is nothing but the texture of the space-time continuum.

Some may dissent from Einstein's view that this simplicity of pattern is inherent in the world of perception, and may claim that it is due to the way in which our minds perceive. A thoroughgoing Kantian would argue that our minds act as lawgivers to nature, prescribing to the external world the ways in which its phenomena shall be perceived by us. The fact that only unbent pennies are found in an automatic machine does not prove that the outer world consists of unbent pennies, but merely that the machine has a selective mechanism which will only accept unbent pennies. In this same way our minds may have a selective action for simple mathematical laws.

On such a view our supposed laws of nature become a mere specification of our own mental processes, telling us little or possibly nothing about nature, but certainly something about ourselves. Yet, if so, what precisely do they tell us? That our minds run naturally and inevitably to matrices, tensors, four-dimensional geometry, and all the various square roots of minus one? Every schoolboy will dismiss such a suggestion as grotesque, and the physicist will certainly concur. If our minds had been thrusting mathematical properties on to nature, we should have designed a more readily intelligible nature than that described in the present book; we may feel just as sure that the repellently difficult matrices and tensors and the brain-racking constructions of four-dimensional geometry come to us from the external world, as the child is sure about the pin which runs into its finger. And if this is so, the same must be true of the simplicity of arrangement of events in the continuum.

Moreover, if the mathematicians merely impress their own mathematical laws on to nature, why cannot the artist, the poet or the moralist do the same and meet with equal success? Why is not the artist able to say—"the sunset will now turn a little more green, or purple; this is necessary to keep it quite perfect as the light decreases" or "the star will appear at the centre of the crescent formed by the new moon, for this is the most aesthetic arrangement of a star and a crescent"? We know that such predictions are worthless. The cloud on the western horizon does not produce the sunset hues by conforming to the canons of art, but by moving in accordance with certain concepts of pure mathematics, and the only way to discover the future of the sunset is to solve the mathematical problem of finding which order of events makes an interval in the continuum continually a minimum.

Finally, if our mathematical minds mould nature to their own laws now, why did they not do so before the twentieth century? It can hardly be supposed that the inherent qualities of the human mind underwent a revolutionary change when Planck published his famous paper in 1900. If the new knowledge expresses a property of the human mind rather than of nature, surely some learned metaphysician might have foreseen that only a mathematical picture could ever be successful, and in so doing have saved science all the misguided effort of trying to draw pictures of other kinds. For three centuries science had projected mechanical ideas on to nature, and made havoc of a large part of nature by so doing.

Twentieth-century science, projecting the ideas of pure mathematics on to nature, finds that they fit as perfectly, and as uniquely, as Cinderella's slipper fitted her foot. We can hardly explain this away by saying that we have merely shaped the foot to fit the slipper, for so many other slippers were tried first and no amount of ingenuity could get the foot into them.

The fact that the mathematical picture fits nature must, I think, be conceded to be a new discovery of science, embodying new knowledge of nature such as could not have been predicted by any sort of general argument. If we could translate our knowledge from the language of phenomena into the language of reality, the word "mathematical" would, I think, have some sort of translation in the latter language; it would not drop away as having represented a mere form of apprehending phenomena. And if this is so, it would seem to suggest that reality must have something of a mental nature about it.

The Road to Ultimate Reality

Yet the fact that the search for a physical reality underlying the mathematical description of nature has so far failed does not of course imply that the search must for ever fail. We must admit it as conceivable that the further advances of science may yet clothe our present mathematical abstractions in new dresses of physical reality, and possibly even of material substance. It is not easy to imagine how formulae in which $\sqrt{-1}$ plays such a prominent part can admit of such an interpretation, yet with the surprising and kaleidoscopic changes of recent years still fresh in our minds, we cannot disregard the possibility. It is, however, so far out of the range of our vision at the present moment, that it is idle to speculate as to what the new dress may be. It may perchance restore mechanical properties to nature, space-time may prove to be a real substantial island floating in something which is not space-time, and so on—nothing can be ruled out as impossible.

Or it may be that no such substantial or material dress will ever be found, and that our knowledge of the universe will for ever remain similar in kind to our present knowledge, a knowledge of our perceptions expressed as a group of mathematical formulae stamped with the stamp of the pure mathematician—the kind of formulae which result from the operation of thought working within its own sphere. In such an event, there may or may not be a non-mental reality behind the form; if there is, it will be beyond our scientific capacity to imagine.

All these possibilities are in the field, since all refer to the future and the unknown. Our positive knowledge of the road along which science is travelling is confined to that which lies behind it. We cannot say how much farther, if at all, the road extends in front, or what the far end of it is like; at best we can only guess.

Some may think that the most plausible conjecture is that the end of the road will be like what is at the half-way house, or perhaps more so. We have already described recent progress in physical science as resulting from a continuous emancipation from the purely human point of view. Our last impression of nature, before we began to take our human spectacles off, was of an ocean of mechanism surrounding us on all sides. As we gradually discard our spectacles, we see mechanical concepts continually giving place to mental. If from the nature of things we can never discard them entirely, we may yet conjecture that the effect of doing so would be the total disappearance of matter and mechanism, mind reigning supreme and alone.

Others may think it more likely that the pendulum will swing back in time.

Broadly speaking, the two conjectures are those of the idealist and realist—or, if we prefer, the mentalist and materialist—views of nature. So far the pendulum shews no signs of swinging back, and the law and order which we find in the universe are most easily described—and also, I think, most easily explained—in the language of idealism. Thus, subject to the reservations already mentioned, we may say that present-day science is favourable to idealism. In brief, idealism has always maintained that, as the beginning of the road by which we explore nature is mental, the chances are that the end also will be mental. To this present-day science adds that, at the farthest point she has so far reached, much, and possibly all, that was not mental has disappeared, and nothing new has come in that is not mental. Yet who shall say what we may find awaiting us round the next corner?

INDEX

α-particles, 148
Absolute time and space, 94 ff., 109,
    142, 174, 176
  velocity, 94
Action, 124, 129
  at a distance, 112 ff., 119
  principle of least, 124, 126
Activities of life, 275, 276
Alkali metals, spectra of, 251
Analysis of light, 20, 28, 31
Animism, 33, 226
Anthropomorphic views of nature,
    33, 43, 226
Aristarchus of Samos, 48
Aristotle, 124, 229
Atomic Physics, 52
Atomism, 15 ff.
Atoms, 15, 16, 182 ff., 243
  nucleus of, 17
  spectra of, 168 ff.
  structure of, 17, 182, 246

β-particles, 148
Berkeley (Bishop), 15, 282, 290
Black-body radiation, 151, 157, 158
Bohr, N., 5, 53, 54, 171, 243, 257
  correspondence principle, 183
  quantum-restrictions, 170, 182,
    192, 245
  theory of spectra, 53, 54, 169, 171,
    177, 204, 244
Born, M., 194, 220
Boyle, R., 292
Bradley, F. H., 4, 40, 68, 110, 145

γ-radiation, 156
Causality, 36, 229, 230, 258, 280
Change, meaning of, 108 ff.
Common-sense, 42, 115
  view of nature, 1, 42, 115, 230
Compton, A. H., 154
Continuum, 100, 101, 293
  curvature of, 118, 129, 131 ff., 293

Copernicus, 48, 49
Cornford, F. M., 73
Corpuscular theory of light, 23,
    24 ff., 83, 123, 160, 227, 288
Correspondence principle of Bohr,
    183
Cosmic Radiation, 156
Coulomb's Law, 101
Curvature, of continuum, 118, 129,
    131 ff., 293
  of space, 134, 136

Dalton, J., 16
de Broglie, L., 194 ff., 204
  waves, 204, 207 ff.
Democritus, 15
Demon, sorting, 272, 276, 279
Descartes, causality, 39
  mind and matter, 38, 75, 282
  primary qualities, 14, 75, 284
  space and ether, 75, 76, 97
  vision, 11
de Sitter, W., 139
Determinism, 36, 44, 228, 254 ff.,
    277
Differential co-efficient, 194
Dirac, P. A. M., 44, 56, 186, 219

Eddington, A. S., 138
Ehrenfest, P., 212, 239, 240
Einstein, A., 5, 43, 56, 65, 84, 93, 94,
    97, 119, 146, 228, 240, 280, 284
  determinism, 228, 281
  free-will, 281
  gravitation, 118 ff.
  quanta, 153, 240
  radiation, 228
  relativity, 3, 50, 51, 84, 93, 101,
    118, 139
  unitary field theory, 127
  universe, 129 ff., 139, 146
Electron, 17, 21, 148, 149, 287, 289
  frequency of, 201, 208

Electron, polarisation of, 251
  spinning, 251
  wave properties of, 54, 66, 200, 208, 212, 222, 237, 242, 249, 286
Entropy, 265 ff.
Ether, luminiferous, 76, 77, 173, 176, 243
Euclid, 120, 128
Events, 11, 102, 262 ff., 286
Evolution, astronomical, 139
  in nature, 108, 144
Expanding Universe, 131 ff., 293

Faraday, M., 84
Fermat's principle, 121
Fitzgerald, G. F., 89, 90
Fitzgerald-Lorentz contraction, 89
Force (mechanical), 35
Franck and Hertz, 53, 169, 228, 243
Free vibrations, 163, 165, 167, 191, 288
Free-will, 36, 279 ff.
Frequency, 152
Fresnel, 23

Galileo, 34, 124, 192
Gamow, 246
Gases, theory of, 16, 147, 156, 157
Generalised theory of relativity, 50 ff., 118
Gerlach, Stern and, 252
Gilbert, W. S., 90
Goudsmit, 251
Gravitation (Einstein), 50, 51, 118
  (Newton), 49, 51, 101, 117
Greek science, 34, 48, 73, 75
Greek views of nature, animism, 34
  atomism, 15
  time and space, 73

Harmonics, 164, 168
Heisenberg, W., 3, 5, 43, 171 ff., 194, 197, 204, 258, 284
Helium atom, 17, 54, 148
Hero of Alexandria, 120, 128, 291
Hertz (Franck and), 53, 169, 228, 243
Hesiod, 74

Homer, 34, 73, 74
Huygens, 35
Hydrogen atom, spectrum, 52, 54, 168, 182
  structure, 17, 52, 243, 244
  structure (wave-mechanics), 243, 246

Idealism, 15, 68, 298
Intensity of radiation, 157
Interference (of light), 121, 216
Interval, in continuum, 102, 105
  principle of least, 126, 295
Indeterminacy, 231 ff., 258
  Heisenberg's principle of, 233, 236, 287
Indeterminism, 228, 230, 258

Jowett, B., 73

Kant, I., causality, 229
  epistemology, 294
  space, 73, 97, 130
Kelvin, Lord, 61
Kepler, J., 49
Kinetic theory, of gases, 16, 147, 156, 157
  of radiation, 156

Laplace, P. S., 58
Lavoisier, 16
Leibnitz, 38, 289
Lemaître, G., 131, 138, 146
Leverrier, U. J. J., 50
Life, activities of, 275, 276
Light, nature of, 19, 22, 24 ff., 63, 66, 83, 242
  particle-theory of, 23, 24 ff., 83, 123, 191, 227
  quanta of Einstein, 153
  reflection of, 18, 120
  undulatory theory of, 19, 23 ff., 63, 83, 121, 123, 161, 242
Local time of Lorentz, 91, 93, 94
Locke, primary qualities, 14, 284
Lorentz, H. A., 86, 89, 90, 93, 101
Lorentz-transformation, 86, 92, 97, 205

Magnetic Induction, 87, 92

Materialism, 32, 64, 261

Matrix (in mathematics), 179 ff., 183, 294

Matter, 11, 12, 113, 289
primary qualities of, 13, 285
reality of, 289
secondary qualities of, 13
structure of, 16 ff., 113, 147

Maupertius, 124

Maximum Entropy, states of, 267, 274

Maxwell, J. C., electromagnetic theory, 84, 87, 254
sorting demon of, 272, 276, 279
theory of gases, 16, 147, 156, 157

Mechanical views of nature, 35, 43, 64, 172, 192, 295, 297

Mercator projection, 115 ff., 129

Mercury, motion of planets, 50

Michelson-Morley experiment, 80 ff., 93, 141

Mind-body bridge, 12, 281

Minkowski, H., 97, 99, 101

Molecule, 16

Monochromatic light, 20

Morley (*see* Michelson)

Nature, mechanical views of, 35, 43, 64, 172, 192, 295, 297
objective, 1, 4, 67, 250, 284, 286
subjective, 2, 4, 65, 68, 284, 285
uniformity of, 7, 36, 230

Nebular motions, 137 ff.

Newton, Isaac, 44, 97, 120, 227
astronomy, 35, 49
determinism, 227, 229
gravitation, 49, 51, 101, 117
mechanics, 14, 35, 44, 124, 191, 192
optics, 23, 227
relativity, 77, 85, 89

Observables and Unobservables, 171 ff.. 284, 285

Oscillations of a solid, 190

Particle picture, indeterminacy of, 257, 258

Particle picture, of electron, 252, 286, 288
of radiation, 22, 123, 160, 167

Pearson, Karl, 56

Phase of vibrations, 122, 181

Philosophical theories, idealist, 15, 68, 298
materialist, 32, 64, 261
mentalist, 15, 298
objective, 1, 68, 284
realist, 69, 298
subjective, 2, 68, 284

Photography, 152

Photon, 26, 52, 83, 153, 175, 216, 288
frequency and wave-length of, 211
waves of, 211, 215 ff., 239, 242

Planck, M., 151, 280, 295
constant of, 191, 233, 234
determinism, 280
free-will, 280
law of radiation, 157
quanta, 151

Plato, change, 109, 110
space, 74, 142, 146
time, 110

Poincaré, H., 43

Poisson-brackets, 186

Polarisation, of electrons, 251
of radiation, 159, 251

Primary qualities, 13, 14, 15, 21, 75, 284

Probability, 224, 253
objective, 226
subjective, 225, 238
waves of, 62, 220, 237, 253

Proton, 17, 21
frequency of, 202, 208
wave properties of, 54, 66, 199, 208

Pythagoras, 48

Quantum theory, 52, 54, 151, 191, 258, 280, 281

Quotations:
Bradley, F. H., 5, 40, 68, 110, 145
Cornford, F. M., 73
Descartes, 39

Quotations (cont.):
    Dirac, P. A. M., 44, 56
    Einstein, A., 56, 229, 280, 290
    Heisenberg, W., 3, 258
    Huygens, 35
    Jowett, B., 73
    Newton, I., 44, 77, 85
    Pearson, Karl, 56
    Planck, M., 280
    Plato, 74, 146
    Russell, Bertrand, 290
    Schopenhauer, 281
    Sidgwick, H., 97
    Whitehead, A. N., 41
    Weyl, H., 229, 281, 291

Radiation, 26, 150, 155
    black-body, 151, 157, 158, 228, 257
Radioactivity, 227, 229, 246, 257
Radium, 228, 229
Rainbow, analogy of, 2, 109
    mechanism of, 3, 13, 21, 47
Random walk, problem of the, 256
Realism, 69, 298
Reflection of light, 19, 120, 151
Relativity, 1, 3, 15, 50, 93 ff., 287
    generalised (gravitation), 50, 51,
        118
    mass, momentum and energy, 153,
        200, 210
    Newtonian, 77, 85, 89
    restricted, 93 ff., 118
Retina of eye, 28, 71
Ritz, principle of, 168, 170
Rotations of atoms and molecules,
    191
Russell, Bertrand, 290
Rutherford, Lord, 227

Schopenhauer, 281
Schrödinger, E., 194, 204
    equation of, 204, 218, 219, 222,
        239, 257
    equation of (atom), 243
    equation of (particle), 248
Secondary qualities, 13
Sensation, threshold of, 231
Sense-impressions, 6, 11, 21, 278

Senses, operation of, 8, 21, 27, 29
Shadows, 23
Soddy, F., 227
Sorting demon of Maxwell, 272 ff.
Space, 96, 141, 215, 292
    curvature of, 114
    Euclidean, 114
    rudimentary views of, 70, 96, 215
Space-like intervals, 105
Space-time, 101, 252
    representation in, 253
Spectra, atomic, 52, 168 ff., 182
Spectroscope, 20, 31
Spinning electron, 251
Spontaneous disintegration (radio-
    active), 227
Statistical atom, 185, 189
Stefan's law, 158
Stern and Gerlach, 252
Sunlight, nature of, 18, 23, 27

Temperature radiation, 151, 157,
    158, 228, 257
Thermodynamics, 262 ff.
Thomson, G. P., 198
Thomson, J. J., 148
Threshold of sensation, 231
Time-like interval, 105
Time, rudimentary views of, 70 ff. 96
    synchronisation, 78

Uhlenbeck, 251
Uncertainty Principle of Heisen-
    berg, 233, 236, 287
Undulatory theory of light, 19, 23 ff.,
    63, 83, 121, 123, 161, 242
Uniformity of Nature, 7, 36, 230
Unitary field theory (Einstein), 127
Universe (Einstein), 129 ff., 136, 146
    expanding, 131 ff., 293
Unobservables, 171 ff., 176, 284, 285

Vision, act of, 11, 28

Wave-equation, of electron, 203,
    219, 248
    of photon, 215
Wave-length, 20, 26

Wave-length, of electron, 198 ff.
　of photon, 215
Wave-mechanics, 54, 194 ff.
Wave-packet, 206, 243, 249
Wave picture, of particles, 194 ff.,
　236 ff., 248, 285
　of photons, 211, 215, 239, 242
Weyl, H., 229, 281, 291
Whitehead, A. N., 41

White light, 18
Wiener, N., 194
Wireless transmission, 26, 31, 79
World-line, 102, 287

X-radiation, 152, 156

Zeeman effect, 251
Zero interval, 106

PRINTED BY WALTER LEWIS, M.A., AT THE UNIVERSITY PRESS, CAMBRIDGE